# LOGICS
## *OF THE*
# KINGDOM
# DYNAMICS

BRIDGING SCIENCE
AND LOGIC WITH
THE KINGDOM OF GOD

# HEATHER
# FORTINBERRY

*Logics of the Kingdom Dynamics: Bridging Science and Logic with the Kingdom of God*
© 2020 by Heather Fortinberry. All rights reserved.

No part of this book may be reproduced without written permission from the publisher or copyright holder, nor may any part of this book be transmitted in any form or by any means electronic, mechanical, photocopying, recording, or other, without prior written permission from the publisher or copyright holder.

All additions or emphasis in scripture quotations are author's own.

Unless otherwise noted, all scripture quotations are taken from the King James Version of the Bible. Public domain in the USA.

Scripture quotations marked (NKJV) are from the New King James Version®. Copyright © 1982 by Thomas Nelson. Used by permission. All rights reserved.

ISBN: 978-1-7352587-0-6 (Print)

978-1-732587-1-3 (Digital)

Printed in the United States of America.

# Contents

| | |
|---|---|
| FOREWORD | 5 |
| PHYSICS AND PHENOMENA | 7 |
| AN ELECTROMAGNETIC CONNECTION | 19 |
| DRAWING GRACE | 23 |
| LET THE WEAK SAY, "I AM STRONG" | 27 |
| RESISTING GRAVITY | 31 |
| PATHS OF RESISTANCE | 35 |
| FEARLESS POTENTIAL | 39 |
| A RIGHTEOUS THEOREM | 43 |
| WAVEFUNCTION COLLAPSE | 45 |
| MAY I HAVE THIS…INTERFERENCE? | 49 |
| REFLECTIONS | 53 |
| STOP! I WANT TO GET OFF! | 57 |
| THE GRAPHITE PRINCIPLE | 59 |
| OUR PRO-VISION | 65 |
| THE APPOINTED TIME | 67 |
| DEVELOPABLE FAITH | 73 |
| WHERE THERE'S VISION, THERE'S LIFE! | 77 |
| CONCEIVE—BELIEVE—RECEIVE! | 79 |
| EVIDENCE | 83 |
| BECOMING FRUITFUL | 87 |
| FOOTPRINTS | 91 |
| "BE-GOTTEN" | 95 |

*LOGICS OF THE KINGDOM DYNAMICS*

| | |
|---|---|
| ABSOLUTE TRUTH | 101 |
| THE COVENANT BOND | 105 |
| IF YOU LOVE SOMEBODY, SET THEM FREE | 107 |
| NUCLEAR FORGIVENESS | 109 |
| THE RECORD BEARER | 115 |
| ENTERING INTO REST | 117 |
| "MY POLICY IS THE SEPARATION BETWEEN SPIRIT AND SILICON" | 121 |
| THE PENDULUM | 125 |
| SEEKING TO DEVOUR | 129 |
| MAN VERSUS MEN | 133 |
| AUTHOR'S NOTE | 139 |
| INDEX | 141 |
| ABOUT THE AUTHOR | 143 |

# FOREWORD

In the twinkling of an eye, the purpose and meaning of one's life can change irreversibly. While a car accident in 1983 was one such event with life-altering repercussions, nothing can compare to the moment when bridging science and scripture became an all-encompassing endeavor.

Having serendipitously looked upon the science section in a Barnes & Noble store, and having a flutter of excitement in response, I knew in my heart that I had landed upon buried treasure, as precious as gold. And I was right! Just *"A*s Moses was divinely instructed when he was about to make the tabernacle, for He (God) said, 'See that you make all things according to the pattern shown you on the mountain'"* (Hebrews 8:5 NKJV), everything in this world is patterned after spiritual truths in heavenly places in Christ Jesus. While the connections may not be obvious, the principles that govern physics, chemistry, and material science, as well as the fundamentals of the four known forces of the universe, are intrinsically woven into the fabric of God's kingdom as well as in Scripture.

As a child, I wanted to be two things: A writer and an archeologist. While brushing dirt from relics never materialized in my life, from 2009 until 2012, when I wasn't teaching psychology, I was taking copious notes about everything that related to physics, chemistry, material science, mathematics, and the like. I was digging deep in unfamiliar ground with the hopes of illuminating bridges that would connect the world of science with the Word of God as written in the Bible—and I *was* finding treasure!

## LOGICS OF THE KINGDOM DYNAMICS

Having uncovered countless connections between natural laws and phenomena and the Word of God, I compiled enough evidence to verify that everything in this natural world does indeed reflect a spiritual or heavenly one. In 2013 I wrote *Logics of the Kingdom: A Scientific Analysis of the Word of God* for those who are scientifically minded, taking an in-depth look at these connections. However, with its applicability to everyday life appearing limited, this book was conceived. As a compilation of thirty-two informal treatises, the fundamental dynamics of the theory, *Logics of the Kingdom* (LOTK) are revealed. Everything in this physical world is discernable according to spiritual principles, and these serve as a bridge between what scientific researchers have uncovered and what the Bible declares as truth.

# PHYSICS AND PHENOMENA

A s man has been created in the likeness of God, "his" mind has a natural affinity to the patterns that God has established. Accordingly, every relationship and formula that has been devised by researchers and scientists to measure and explain natural phenomena can likewise be used to measure and explain spiritual phenomena. A truth exists: Everything on this earth—as well as above the earth, under the earth, and in the oceans—adheres to a heavenly pattern. This life is made up of many dimensions, which are all connected, in that they are governed by the same fractal-like patterns that can be recognized at ever increasing or diminishing scales. The importance of earth's reflection of heavenly patterns is underscored in Hebrews 8:5:

> As Moses was admonished of God when he was about
> to make the tabernacle: for, See, saith he that thou make
> all things according to the pattern shewed to thee in
> the mount.

Why was God so adamant that the pattern showed to Moses be replicated precisely? Because this was the building where the earthly high priests would enter in to the Holy of Holies. This was the building where they would offer sacrifices to God for the people, which was to mirror the office to which Jesus had been pre-ordained. Jesus, who was the Word in fleshly form, came to earth to make one last sacrifice for all humanity, and "He (has) obtained a more excellent ministry (than the earthly priests), by how much also he is the mediator of a better covenant, which was established upon better promises" (Hebrews 8: 6). We are heirs of better promises. The Bible has been left to us as a gift, so

# LOGICS OF THE KINGDOM DYNAMICS

that we may search out these great and precious promises for ourselves, and everything in this earth has been made after a pre-ordained pattern in heaven.

Let's begin at the beginning: "In the beginning God created the heavens and the earth" (Genesis 1:1). As this is the first sentence of the Bible, one thing stands out above all: There is no speculation regarding who created the heavens and the earth! It was God! And who is God? Speaking to Moses, this entity specified who He was:

> And God said to Moses, "I AM WHO I AM." And He said, "Thus you shall say to the children of Israel, I AM has sent me to you" (Exodus 3:14 NKJV).

The heavenly pattern of God's identity as the "I AM" has as its reflection the substance of DNA. Not only is DNA present in nearly every living organism and is an identifier, but it also is a chief provider of instructional information. As God the Father sent His commands to Jesus, so, too, is the process of DNA sending its message to messenger RNA. Verifying this relationship, Jesus said to his disciples, "Peace to you! As the Father has sent Me, I also send you," (John 20:21 NKJV). Messenger RNA then acts as a mediator between a DNA template strand and protein, where the message is read and eventually translated so that protein can go out and nourish the body. As it is in the natural, so it appears in the spiritual, for 1 Timothy 2:5 says, "For there is one God and one Mediator between God and men, the Man Jesus Christ" (1 Timothy 2:5 NKJV). And as the "Word was made flesh and dwelt among us, (and we beheld His glory as the only begotten of the Father,) full of grace and truth" (John 1:14), we know that Jesus and the Word are the same. That its job is to nourish is also supported in Matthew 4:4 which says, "Man shall not live by bread alone, but by every word that proceeds from the mouth of God."

There is also the RNA World Theory, which counters the assumption that DNA was the first life-building molecule. Remaining in the nucleus, DNA needs messenger RNA to cross the barrier membrane and take its message to be translated into protein. Making the case for RNA coming before DNA, this theory highlights the fact that because protein is required to build life-sustaining matter, RNA had to have come first. While RNA is built from a DNA template strand, because of the interchangeability of information between the two molecules, it is not beyond reason to suggest that RNA came first. And if the true identity of messenger RNA would be the Word of God, John 1:1 (NKJV) would be supported, which says, "In the beginning was the Word and the Word was with God, and the Word was God." Intricately tied together, God and His Word as well as DNA and RNA both appear to have initiated this world's matter.

Jesus is also referred to as the Light: "In him was life; and the life was the light of men" (John 1:4). As any believer can attest, there is a lightness of being that is often experienced, a warmth from within or a light felt emanating through one's eyes, when Jesus is accepted into the heart. It might also be understood that the Light is the conscience that can discern between good and evil. He is also the Word incarnate:

> In the beginning was the Word, and the Word was with God, and the Word was God. He was in the beginning with God. All things were made through Him, and without Him nothing was made that was made (John 1:1-3 NKJV).

These four things: DNA, RNA, Light, and the Word, can be substituted for God and His Son, Jesus Christ, in gaining insight and understanding into God's identity, form, and function.

Jesus sent forth life-creating light into the world every time He spoke. His Word behaved as a seed, which when spoken, went forth searching for an understanding heart in which it could be planted. "But he who received seed on the good ground is he who hears the word and understands it, who indeed bears fruit and produces some a hundredfold, some sixty, some thirty" (Matthew 13:23 NKJV). Existing within the seed of the Word of God is life-creating light. This light can be understood as the Witness spoken about in 1 John 5:7-8:

# LOGICS OF THE KINGDOM DYNAMICS

> For there are three that bear witness in heaven, the Father,
> the Word, and the Holy Spirit; and these three are one.
> And there are three that bear witness on earth: the Spirit,
> the water, and the blood; and these three agree as one.

This light is in both the Word and the Spirit, and can be understood as revelation knowledge, which allowed Peter to discern Jesus's identity when He asked him, " 'But who do you say that I am?' Simon Peter answered and said, 'You are the Christ, the Son of the living God.' Jesus answered and said to him, 'Blessed are you Simon Bar-Jonah, for flesh and blood has not revealed this to you, but My Father who is in heaven'" (Matthew 16:15-17). Now, as the "faithful witness" *(Revelation 1:5)* Jesus "was that true Light which gives light to every man coming into the world" (John 1:9).

As a witness of God the Father's plan for His kingdom, in revealing to His disciples why He spoke in parables, Jesus said:

> To you it has been given to know the mysteries of the
> kingdom of God, but to the rest it is given in parables,
> that "Seeing they may not see, and hearing they may not
> understand." Now the parable is this: The seed is the
> Word of God (Luke 8:10-11).

Within the Word of God, therefore, is the light of revelation knowledge. And that light can only be received or reflected by a light of a similar frequency. When an atom is struck by an incoming photon, an electron of a similar energy to the photon will receive it and absorb its light. In response, it will jump to the next higher level, and on its descent back down it will emit the excess energy in the form of light. This provides the natural pattern of its spiritual counterpart that Matthew 5:16 speaks about: "Let your light so shine before men that they may see your good works and glorify your Father in heaven."

This light that emanates from men upon receipt of incoming light (thoughts) exists on a continuum between total reflection and total absorption. Matthew 6:22-23 (NKJV) speaks of this light:

> The lamp of the body is the eye. If therefore your eye is good, your whole body will be full of light. But if your eye is bad, your whole body will be full of darkness. If therefore the light that is in you is darkness, how great is that darkness!

The greater the absorption of others' light, the darker one's eye may be. Whereas when one reflects—or rather connects—with others' light, this person's eye, or light, is valued as good. These light frequencies that span from reflective white light to absorbing blackness interact with each other according to wave dynamics, where constructive and destructive interference can account for most daily interactions. We can therefore discern the light of others through how our own light reacts in response. We can become excited because our waves are on the same frequency and in the same phase, or we can become withdrawn. The reflective light can also be understood as love, while the absorbing dark light can be understood as fear. First John 4:18 mentions this relationship: "There is no fear in love; but perfect love casts out fear, because fear involves torment." And 1 Corinthians 13:8 says, "Love never fails." Together these scriptures articulate the dominion of the white light of love over the dark light of fear.

God's purpose is sure! As "God is love" (1 John 4:8), "He gave His only begotten Son, that whoever believes in Him should not perish but have everlasting life" (John 3:16 NKJV). Connected closely to this was Jesus's purpose, which was to fulfill all righteousness. We see this play out when Jesus approached John the Baptist at the Jordan River to be baptized by him.

> And John tried to prevent Him, saying, "I need to be baptized by You, and are You coming to Me?" But Jesus answered and said to him, "Permit it to be so now, for thus it is fitting for us to fulfill all righteousness" (Matthew 3:14-15 NKJV).

## LOGICS OF THE KINGDOM DYNAMICS

The plan of redemption for everyone is revealed in Mark 16:15-16 (NKJV), and as a man born into this world, Jesus had to fulfill His own command to His disciples: "Go into all the world and preach the gospel to every creature. He who believes and IS BAPTIZED will be saved; but he who does not believe will be condemned." Being a "preacher of righteousness" (2 Peter 2:5), Jesus had to fulfill every lawful obligation, as well as all prophetic utterances according to His declaration in Matthew 5:17-18:

> Do not think that I came to destroy the Law or the Prophets. I did not come to destroy but to fulfill. For assuredly, I say to you, till heaven and earth pass away, one jot or one tittle will by no means pass from the law till all is fulfilled.

From this we understand that love and God are interchangeable, and while righteousness includes following the law of God (love), belief in the gospel is at the center.

> For I am not ashamed of the gospel of Christ, for it is the power of God to salvation for everyone who believes, for the Jew first, and also for the Greek. For in it the righteousness of God is revealed from faith to faith; as it is written, "The just shall live by faith" (Romans 1:16-17 NKJV).

In addition to Love, Light, the Word, DNA, and RNA, there are the scientific formulas that have been used for centuries to measure the world's phenomena. From Newton's Laws to Einstein's Relativity, these have remained limited to providing quantitative measurements to the physically viable "things" in this universe. In reality, scientific formulas appear to have qualitative counterparts that can be used to accurately measure spiritual things such as the fruit of the Spirit, and they have just been waiting to be revealed!

So let us begin with work, which is a preoccupation with most of us, and whose formula is WORK = FORCE x DISTANCE. Here, the scientific formula describes how work may be measured, but if we are to understand this in spiritual terms, we must equate these terms to spiritual phenomena, which under normal circumstances are not readily quantifiable. But as the Word is truth, and truth can be backed by physical evidence, there will be a scriptural reference to the phenomena, just as sure as "Till heaven and earth pass, one jot or one tittle shall in no wise pass from the law, till all be fulfilled" (Matthew 5:18 NKJV). For instance, "I must work the works of Him that sent me, while it is day; the night cometh, when no man can work. As long as I am in the world, I am the light of the world" (John 9:4-5). The "work" in this context should be able to be verified as being a product of the spiritually-equivalent force and distance.

One definition of *force* is "to break open or through," so we can readily identify faith as a force. In Mark 2:4-5 we see four men break through a roof in order to lower a paralyzed man down to Jesus:

> And when they could not come near Him because of the crowd, they uncovered the roof where He was. So when they had broken through, they let down the bed on which the paralytic was lying. When Jesus saw their faith, He said to the paralytic, "Son, your sins are forgiven you."

Their act of breaking through the roof was a display of their faith, for they believed that the man would be healed if they could get him to Jesus. And he was healed! But not before Jesus was accused of blasphemy. But he said to his accusers,

> "Which is easier, to say to the paralytic, 'Your sins are forgiven you,' or to say, 'Arise, take up your bed and walk'? But that you may know that the Son of Man has power on earth to forgive sins"—He said to the paralytic, "I say to you, arise, take up your bed, and go to your house." Immediately he arose, took up the bed, and went out in the presence of them all (Mark 2:9-12 NKJV).

Therefore, Work = Faith x Distance.

## LOGICS OF THE KINGDOM DYNAMICS

We can, therefore, validate the involvement of faith in work, and in addition, John 6:29 confirms this: "This is the work of God, that ye **believe** on Him whom He hath sent." So in a spiritual sense, *work* relates to having faith, a force which can be substantiated in that faith is used to break open the blessings of God or break through the obstacles presented before us.

Now, if force equals faith in the equation for work, what spiritual phenomena can be substituted for distance? What do we require to "go the distance?" Aside from faith that we can achieve the prize, we need faith's fruitful companion, patience, which is able to weather the storms and endure hardship with faith by its side.

These "power twins" are referenced in Hebrews 6:18:

> That by two immutable things, in which it was impossible for God to lie, we might have a strong consolation, who have fled for refuge to lay hold upon the hope set before us.

Now while this scripture doesn't spell out that it's referring to faith and patience, if we scroll up to verses 11-12, it becomes clear, "And we desire that every one of you do shew the same diligence to the full assurance of hope unto the end: (this is the hope to which verse 18 alludes,) That ye be not slothful, but followers of them who through **faith** and **patience** inherit the promises." As God is faithful to accomplish what He pleases, and that it will prosper in the thing for which He sent it (see Isaiah 55:11), when we're followers of His Word and apply our own patience, we ourselves have the opportunity to bear fruit, because His Word is truth.

So WORK = FAITH x PATIENCE, with *force* equating *faith*, and *distance* equating *patience*. Now, A quantity related to these is POWER = WORK/TIME. With *work* as FAITH x PATIENCE, to find an equation that lets us evaluate the power in any situation or person, we only need to reconcile the notion of time. Well, according to the natural phenomenon of time, it is considered a one-way arrow toward decay or entropy. When

I was trying to reconcile this equation with God's kingdom's principles, I was walking my dog, Willie, around the corner and down the hill. "If time signifies decay and You are not a God of decay, how can I reconcile the notion of time?" I asked the Lord. Instantly I heard His reply: *"The notion of time does not exist with ME, but must be replaced with the renewing, restoring, and resurrecting power of My Word."* Therefore, measurements relating to power, force, distance, work, and time relate to faith, patience and the renewing, restorative, and resurrecting power of God that enables a complete reversal of the enemy's plans and purposes in this world.

In addition, in light of the formula for force, which is FORCE = MASS x ACCELERATION, we see faith as a force used to break through plateaus, as it is required to "Press toward the mark for the prize of the high calling of God in Christ Jesus" (Philippians 3:14). And, as verse nine and ten says:

> Be found in Him, not having mine own righteousness, which is of the law, but that which is through the faith of Christ, the righteousness which is of God by faith: That I may know Him and the power of his resurrection.

While in many instances, *mass* can more appropriately be substituted for weight, in this case it may be substituted for substance, as in the "substance of things hoped for" (Hebrews 11:1). And *acceleration* can represent an increasing change in the velocity of our application of faith, as in Romans 1:16-17, which says:

> For I am not ashamed of the gospel of Christ: for it is the power of God unto salvation to everyone that believeth; to the Jew first, and also to the Greek. For therein is the righteousness of God revealed from faith to faith: as it is written, "The just shall live by faith."

Our faith should only get stronger as our relationship with the Word gains strength, or as we continue in the Word (see John 8:31),

*LOGICS OF THE KINGDOM DYNAMICS*

for "Faith comes by hearing, and hearing by the Word of God" (Romans 10:17 NKJV).

Measurements, such as length, width, depth and height, which when multiplied, produce the volume of an object, can by extension be understood by Ephesians 3:17-19:

> That Christ may dwell in your hearts by faith; that ye, being rooted and grounded in love, may be able to comprehend with all saints what is the breadth and length and depth and height; and to know the love of Christ, which passes knowledge, that you might be filled with all the fullness of God.

Volume, therefore, seems to point to the unlimited potential (fullness) that is available with God, as Mark 10:27 (NKJV) declares: "With men it is impossible, but not with God: for with God all things are possible." Now, proving this equivalency requires another arbitrarily selected equation involving volume, such as DENSITY = MASS/VOLUME.

When we apply the concept of density to an object or person, what we often land upon is a notion of mental or physical thickness, where light (among other things) can hardly penetrate. With an equivalent of mass being weight, further insight may be gained by reading Hebrews 12:1:

> Wherefore seeing we also are compassed about with so great a cloud of witnesses, let us lay aside every weight, and the sin which does so easily beset us, and let us run with patience the race that is set before us.

Here we see that weight is equivalent to emotional baggage or sin. So DENSITY = EMOTIONAL BAGGAGE or SIN/UNLIMITED POTENTIAL. Now, to verify the applicability of the equation, we must verify the experiential reality if such an equation were applied to people's lives.

Therefore, if the density (thickness) was large, then the emotional baggage and sin would be large as well, and the unlimited potential would be small, such as would appear in a numerical equation: A density of 100 would mean that the emotional baggage or sin was at least 100 as well, with the unlimited potential being closer to 1.

However, if the emotional baggage and sin was 1 we could assume that the weight and sin had been laid aside already, and the expenditure of potential energy, as one could run and potentially win a race, would be close to "100." The resulting number would therefore be a fraction, and the density of the person would be small. This makes sense experientially, so we may affirm that we may use this formula to gain a verified conclusion about a person's state of density—or more specifically, the psychological heaviness or spiritual darkness to which they have succumbed.

The concept of potential energy is interesting, as it becomes one side of a fluid system where potential and kinetic energy continuously transition into one another. In the example of a pendulum, the midpoint of the pendulum's arc marks the point at which kinetic energy transitions into potential energy, and the apex/pinnacle point marks the point at which potential energy transitions into kinetic energy, and the shift from one to the other is continuous. The potential energy in this example is gravitational potential energy, where an object's potential energy depends on its position. The potential energy expended in running a race, however, would depend on the stored chemical energy of the food consumed and the training involved in building muscles. And while training muscles requires resistance against the gravitational pull of the earth, the potential energy that is converted to kinetic energy relies on the chemical ATP (adenosine triphosphate), which is the body's way to store and transport energy.

What is potential energy's end game? All potential energy is stored, so that work can be accomplished! So all potential energy is stored, with the end result that we will live by faith and patience, according to our spiritually-equivalent equation. And as the "Just shall live by faith," (Romans 1:17) we will be among the justified, and if faith and patience operate as they should, they will bring us out of bondage to sin and

*LOGICS OF THE KINGDOM DYNAMICS*

make us free. We know this is true according to Romans 10:17 (NKJV), which says: "Faith comes by hearing and hearing by the Word of God" and "If ye continue in My word, then are ye my disciples indeed; and ye shall know the truth and the truth shall make you free (John 8:31-32).

The Word of God becomes stored in our hearts as we continue to hear and speak it. This stored energy can be understood as light potential, which is able to create whatever the Word communicates when it is released. What is imperative, however, is faith as Hebrews 4:2 declares:

> For unto us was the gospel preached, as well as unto them: but the word preached did not profit them, not being mixed with faith in them that heard it.

The bond formed from hearing the Word of God preached and mixed with a full persuasion that what is heard is truth, is like all other chemical bonds. As chemical bonds are formed, there is a consequent release of heat energy, which causes a compacting of elements within the bond. As in a seed, the compaction of elements releases heat energy, but in exchange for the release is a stored potential energy of life-creating light. As the parable of the sower reveals in Luke 8:11: "Now the parable is this: the seed is the word of God," the Word of God has stored within itself life-generating creative light potential, and when spoken and mixed with faith, it brings to fruition every promise that has been received by an understanding heart.

# AN ELECTROMAGNETIC CONNECTION

There is little doubt (at least in the minds of neuroscientists such as Dr. Caroline Leaf, author of *Who Switched Off My Brain?*) that the thoughts we allow into our minds can wreak havoc on the structural composition of our brains–and consequently on every other facet of our lives. Validated both scientifically as well as experientially, the connection between our thoughts and healthy functioning appears to be undeniable. However, can this knowledge be verified both scientifically as well as scripturally? Proverbs 23:7 (NKJV), which says, "As he thinks in his heart, so is he," although simple, provides a scriptural reference that helps uncover how a fundamental force of the universe is related to the heavenly realm. From this, we gain an understanding that one's thoughts are intricately connected to one's sense of identity; but how our own sense of identity relates to the forces of the universe, must be explored for new insight to break forth.

Preliminarily, we must understand that everything in the earthly and planetary realm reflects a spiritual pattern or process that is revealed in scripture. Matthew 6:9-10 suggests this truth in its proclamation of the beginning of the Lord's Prayer: "Our Father which art in heaven, hallowed be thy name, thy kingdom come. Thy will be done in earth, as it is in heaven." This much is clear: If God's will is done, earth will reflect heaven!

What is God's will? It is His Word, which, according to 1 John 5:7, is one of the record bearers in heaven. What is a record bearer? It is the

## LOGICS OF THE KINGDOM DYNAMICS

*final word* on a subject, much like the Constitution of the United States. Ideally, everything must line up with what is written in that document for justice to prevail. And as God is a just God, we see earthly things reflect those of heaven. From the very beginning, in heavenly places was the Word:

> In the beginning was the Word, and the Word was with God, and the Word was God (John 1:1).
>
> And the Word was made flesh, and dwelt among us, (and we beheld His glory, the glory as of the only begotten of the Father) full of grace and truth (John 1:14).

These scriptures draw a direct connection between the Word—which began in the heavenly realm—and Jesus. As the Word incarnate, Jesus was the direct reflection of that Word in heavenly places. God's will in action, where earthly things reflect heavenly things, is also proclaimed in Hebrews 8:5 when He (God) commanded Moses to build the tabernacle:

> As Moses was admonished of God when he was about to make the tabernacle: for, See, saith he, that thou make all things according to the pattern shewed to thee in the mount.

This indicates that things that are made on earth have already been predesigned in the heavenly realm.

Understanding that everything has already been created in the conceptual realm, we uncover an intriguing fact: Thoughts are the spiritual prototype of life-sustaining light. They can create a reality as well as keep it going, and they are also intricately involved in stabilizing one's identity. When someone's identity or sense of self, which invariably is backed by thoughts, is without question, they are considered stable. However, when someone's identity or sense of self is uncertain or confused, it is assumed that their instability is caused by erratic thought patterns. All we need to do is turn on the television to be reminded that the world is filled with unstable people whose actions have been governed by an apparently unstable brain. Whether drugs, alcohol, mental illness, or extreme ideological indoctrination is involved, all these factors have one thing is common: they affect thought patterns.

Interestingly, it appears that people are no different than every other element: Every element can be identified by two things: The number of protons in its nucleus and its spectral emission. We, too, are intricately tied to these two identifiers. An element's spectral emission relates to the colored light (frequency) that is emitted after excitation, and we find its spiritual significance when we relate it to our own light as referenced in Matthew 5:16:

> Let your light so shine before men, that they may see your
> good works, and glorify your Father which is in heaven.

Thus far, the notion of stability isn't addressed; however, when speaking about an element's proton count defining its identity, the notion of stability is addressed. For an element to be considered stable, the number of electrons that surround its nucleus must be the same as the number of protons within the nucleus. When the nucleus is understood as the mind, then the electrons that must coordinate with the nucleus for there to be stability can rightfully be understood as thoughts. To confirm this association, when a person is instable, their thoughts determine their condition.

While the universal applications of electrons are vast, narrowing in on the fundamental force of electromagnetism, where the electric field and the magnetic field are perpetually perpendicular to one another, one relationship comes to the forefront: These two fields are intimately related and are thereby mutually affected by change. Accordingly, if we relate electrons to thoughts, and the electric field is governed by electrons, then thoughts are governors. Similarly, the magnetic field is also governed by electrons and therefore thoughts, but the charges within this field create the dynamics of attraction and repulsion. These dynamics lead us to identify this field as governing emotions, which are known to govern people's likes as well as dislikes. Interestingly, these

## LOGICS OF THE KINGDOM DYNAMICS

fields are dependent on each other in the same exact way that thoughts and emotions appear to be: When there is a change in one field, there is a subsequent change in the other, just as when our thoughts change about an event, our emotions about the event change accordingly; and when our emotions change about a person, our thoughts also follow suit.

While this interchangeable connection between electrons and thoughts may be a shift in perspectives, it is one to be mindful of. Discouragement, doubt, despair, and despondency are all tools used to keep us in an emotionally weakened state. Why? Because, as 1 Peter 5:8 (NKJV) states, "The devil walks about like a roaring lion, seeking about whom he may devour." If he can get to our emotions, he can get to our thoughts, and vice versa. And if he can govern our thoughts, he can govern our identity, and this is key! In these last days—more than in any other time—we need to hold fast to our identity as heirs of God and His Son Jesus Christ so we can Overcome satan by the blood of the Lamb and by the word of our testimony (see Revelation 12:11).

# DRAWING GRACE

I was listening to a CD a couple of years ago about the favor of God. Faith erupted in my heart as I listened and heard the word, but as I began to meditate on the scriptures regarding grace and recalled what certain ministers were espousing about the subject, I began to see it all in a new light. This particular minister didn't approach grace as a license to sin, as some others were teaching, but in the context of favor, which is ultimately brought about by living a life of obedience to God.

My spirit bore witness with this teaching because my journey with the Lord began with Luke 12:48 (NKJV) ever repeating in my mind: "For everyone to whom much is given, from him much will be required." But I couldn't fully embrace this message that I was listening to about obedience, because the way it was being presented, favor and grace were basically interchangeable. Perhaps prior to a "false grace" message being preached, I wouldn't have thought so deeply about it, however, having realized that a grace message was leading some down a path of error, I asked the Lord to clarify the difference between favor and grace. Surprisingly, He gave me an illustration that mirrors the physics involved with magnetism.

*Favor* is defined as "something done or granted out of goodwill rather than from justice or for rumination." And *grace* is defined as "the free and unmerited favor of God, as manifested in the salvation of sinners and the bestowal of blessings." However, when we dig more deeply into the governing forces of God, a principled dynamic begins to appear. The relationship that God desires with mankind

## LOGICS OF THE KINGDOM DYNAMICS

is covenantal, the force of which is like a bond that cannot be easily broken.

The Lord provided me with understanding: Grace is like the force of a bar magnet that draws other bar magnets toward itself; this "drawing" force has been arbitrarily assigned the value of "negativity." While this force draws things to itself, its behavior is contingent on the force fields emanating out of the objects with which it interacts. For instance, when a smaller bar magnet with the same force yet of a smaller magnitude approaches it, what does it do? It flips the smaller bar magnet around so that the other side (positive) can connect to it, as opposites attract.

With this illustration, I saw that the grace of God is like the negatively-charged force field of a bar magnet. When it is approached by another negatively-charged bar magnet, it causes the smaller one to turn around, which is the definition of *repentance*. This force field, therefore, can be understood as the goodness, or grace, of God that leads to repentance (Romans 2:4). And what happens when a positively-charged force field interacts with this force? Unless it has been turned around by the goodness of God, we may assume that the character of this force field is "pride," and we only need to look at 1 Peter 5:5 to understand what occurs in this dynamic:

> Yea, all of you be subject one to another, and be clothed with humility: for God resists the proud, and giveth grace to the humble.

The connections, or bonds, that occur between God and man require humility, which flows out of knowledge that their works have not afforded them God's mercy and favor. The proud, however, are characterized by their lack of humility. So what happens? When a prideful or arrogant person is in the presence of God, the other side of God's power (the "positive" side, as of a bar magnet) flips around and that "positive" force field runs up against the "positive" force in the person, and as two positive forces meet, there is nothing but resistance, with God's "positive" force likely pushing the smaller away from Him. We see this dynamic playing out as those who resist God are resisted in turn, and this resistance is seen as offense. And I find it interesting that those who resist righteousness appear forever offended—by everything.

# LET THE WEAK SAY, "I AM STRONG"

Joel 3:10 declares, "Beat your plowshares into swords and your pruning hooks into spears: let the weak say, I am strong." Now, while part of this scripture is a command for people to turn their attention toward the battle in front of them, the end provides a clue as to how the battle is going to be won: One must SAY something, and that "something" better be a proclamation of one's strength and not of their weakness. The point here is that no one is a victim to their situation if they SAY and believe something different; we are not bound to our station in life.

While there is a false sense of security in victimhood, as people who feel that nothing can be done, feel free to do nothing, if they were to recognize that their words could turn their lives around, this would be ultimately liberating. Our fundamental identities can undergo a transformation. And that is good news! One of the strongest statements to ever be spoken is "I AM." This, after all, is the true identity of God as He revealed it to Moses:

> And God said to Moses, "I AM WHO I AM." And He said, "Thus you shall say to the children of Israel, 'I AM has sent me to you'" (Exodus 3:14 NKJV).

As an identifier of God Himself, "I AM" has obvious spiritual significance. First, belief in His fundamental existence is profound, for, "He that cometh to God must believe that he is, and that he is a rewarder of them that diligently seek him" (Hebrews 11:6). And while our belief

## LOGICS OF THE KINGDOM DYNAMICS

in His existence as the essence of life is vital, what we allow ourselves to think about ourselves also determines our paths. Remember, as He (Jesus) is in the world, so are we (see 1 John 4:17). Therefore, whatever Jesus is, a faithful witness (Revelation 1:5) for example, can become our own identity as well if we call those things that be not as though they were (see Romans 4:17). Our identity as children of God, for whom the promises of God apply, will solidify our ability to take a stand in faith against all that may come against us.

Every element in this world, from hydrogen to uranium and beyond, belongs to a decay chain, which includes all the elements that it could become, should the element gain or lose protons. Interestingly, while these transformations occur at the quark level, where up quarks turn into down quarks and vice versa, an energy carrier called a "W boson" comes in and provides the energy which enables this change from one element's identity into another. And this is exactly what is needed for us! With the energy carrier of the Word of God and the Holy Spirit, we, too, can transition from an identity of being one without God in their life to one of being a son of God. "But as many as received him, to them gave he power to become the sons of God, even to them that believe on his name" (John 1:12).

Second Corinthians 5:17 (NKJV) also says this:

> Therefore, if anyone is in Christ, he is a new creation; old things have passed away; behold; all things have become new.

Like every other element that undergoes a process of "decay," where protons turn into neutrons and neutrons turn into protons and provides them with a new identity, we can become new as well! While the transition occurs at the quark level for the elements, for us, the transition occurs at the heart level, and our new identity is solidified as children of God. And just as it is essential that a W boson provide the energy for an element's transition, it is essential for the Word of God to provide the light of revelation knowledge, as "He that cometh to God must believe that he is, and that he is a rewarder of them that diligently seek him" (Hebrews 11:6). Our own identity as children of God should enable us to assume an identity of strength, according to Joel 3:10, which says, "Let the weak say, I am strong."

> But what does it say? "The Word is near you, in your mouth and in your heart" (that is, the word of faith which we preach): that if you confess with your mouth the Lord Jesus and believe in your heart that God has raised Him from the dead, you will be saved. For with the heart one believes unto righteousness, and with the mouth confession is made unto salvation (Romans 10:8-10 NKJV).

# RESISTING GRAVITY

As light and dark, cold and heat, particles and anti-particles, and life and death, two sides of every proverbial "coin" have appeared throughout the ages. As one of the fundamental forces, gravity has long been recognized as the force that keeps us grounded as the earth pulls us toward itself. It's interesting to note that the root word of this force is *grave*, as that appears to be the result of people's obedience to this force.

People start out in life with a proclivity toward rebellion, and while most rebellion does not serve a meaningful purpose, one rebellion has served humanity well—the need to stand, walk, run, and climb. All these actions rebel against the force of gravity, or the power of the grave. Like a potential well that traps all who slip and fall into it, the only hope for those trapped is an external force that sweeps in and catapults them out of the well. If the grave or the well were to have a motive, it would be to keep all who were trapped stagnant, unproductive and depressed, so the catapulting force would want them free!

This is the reason, Jesus—the Word incarnate—came. Isaiah 61:1-3 mentions specifically why Jesus came on earth:

> To preach good tidings unto the meek; he hath sent me to bind up the brokenhearted, to proclaim liberty to the captives, and the opening of the prison to them that are bound; to proclaim the acceptable year of the Lord, and the day of vengeance of our God; to comfort all that mourn; to appoint unto them that mourn in Zion, to give unto them beauty for ashes, the oil of joy for mourning,

# LOGICS OF THE KINGDOM DYNAMICS

the garment of praise for the spirit of heaviness; that they might be called trees of righteousness, the planting of the Lord, that he might be glorified.

It is apparent from this scripture that to become a tree of righteousness and to be considered the planting of the Lord that would glorify God, we must rely on a force that is ready and willing to catapult us out of the wells that entrap us. Yes, we must rise up and resist the gravitational pull of forces that pull us down by declaring that the joy of the Lord is our strength (see Nehemiah 8:10), and by believing that this is a promise that we can accept as true. We can also offer up the sacrifice of praise as Hebrews 13:15 suggests: "Let us offer the sacrifice of praise to God continually, that is, the fruit of our lips, giving thanks to his name."

Forces that keep us in the well are opposed to joy and praise as Matthew 6:16-18 (NKJV) alludes:

> Moreover, when you fast, do not be like the hypocrites, with a sad countenance. For they disfigure their faces that they may appear to men to be fasting. Assuredly, I say to you, they have their reward. But you, when you fast, anoint your head and wash your face, so that you do not appear to men to be fasting, but to your Father who sees in the secret place; and your Father who sees in secret will reward you openly.

We must stand tall in faith, believing that God is by our side every step of the way—because He is!

When we were young, we chose to rebel against the force of gravity, and rose and ran. It could be easily stated that we rebelled against the force that would have preferred to hold us down. Every action is thus a disobedience to an opposing force; if we rise, we are rebelling against reclining. If we recline, we are rebelling against rising. All of this has to do with the ultimate choice as Deuteronomy 30:19 declares:

> I call heaven and earth to record this day against you, that I have set before you, life and death, blessing and cursing; therefore choose life, that both thou and thy seed may live.

What we choose to resist is ultimately significant, because an open embrace of its opposite is implied. When we resist death, we embrace life, but if we resist life, then we embrace death and make a covenant with it. Our choice of life or death, and blessing or cursing also has ramifications that will be made known well beyond this life's existence, as first Corinthians 15:52-54 (NKJV) says:

> For the trumpet will sound, and the dead will be raised incorruptible, and we shall be changed. For this corruptible must put on incorruption, and this mortal must put on immortality. So when this corruptible has put on incorruption, and this mortal has put on immortality, then shall be brought to pass the saying that is written: "Death is swallowed up in victory."

Here we see that the dead in Christ shall rise and counter gravity's force; and this is the ultimate description of power: POWER = FAITH x PATIENCE/THE RENEWING, RESTORING, RESURRECTING ABILITY OF GOD.

# PATHS OF RESISTANCE

The whole conceptual framework behind the notion of "Paths of Resistance" holds personal significance. In line with Isaac Newton's second law of motion, which says that for every action there is an equal and opposite reaction, many years of my own life could be characterized by an overall thrust in the forward direction, with a simultaneous engagement of a resistance to a pull in the opposite direction. Having been in a coma, paralyzed, with a quarter of my brain "dead," or inactive, the resistances took on many forms—from resisting self-pity and lethargy, to resisting overtaxing my brain. This latter was a product of an overzealousness born out of the fear of losing any knowledge regained.

While my own journey is not ordinary, the dynamics in pushing in one direction while simultaneously resisting the other direction are universally experienced; after all (as I have found out), all human experience—physical as well as spiritual—does indeed mirror scientifically-verified laws. Every step we make in one direction involves an often-subconscious resistance of the opposite direction; when we get out of bed in the morning and plant our feet on the ground, this action could be framed as a resistance to a force that is pulling us toward itself—a force that wants us to resist getting out of bed. If we stay in bed, we may likewise be said to be resisting the force involved in rising.

Without the expending of energy, an accumulation of potential energy is impossible. When work is done, an equal amount of energy to that which was expended is transferred to the object that was moved.

## LOGICS OF THE KINGDOM DYNAMICS

For example, in the case of a boulder being raised thirty feet, having had work expended to raise it to such a height, the same amount of energy that was required to raise it, now is potentially stored within the boulder. The potential that it now has would be fully realized should it drop, particularly if it fell upon unsuspecting passersby.

Now, if we were to replace "work" with faith x patience according to the LOTK theoretical conversion, we might begin to see how faith and patience becomes stored and accumulates power within those who are exercised thereby. As an example of this power, Jesus answered the woman who had touched His clothes in the crowd, saying, "Daughter, your faith has made you well. Go in peace, and be healed of your affliction" (Mark 5:34 NKJV). Her exertion of faith was stored as potential energy, which brought healing to her body. What may not be evident, however, is the unspoken resistance to her not following Jesus and touching His clothes.

The generation of potential energy, therefore, requires that the force of resistance be in operation on some level. And this has everything to do with us becoming the most authentic and courageous people that we can be—people who can overcome obstacles and fulfill our own God-given potential. When we follow the paths of least resistance, however, we do not generate potential. This begs the question: Which is the larger obstacle standing in the way of personal growth: Fear or laziness? While an argument may be made for either of these holding the title, I might suggest that both are two sides of the same coin, so to speak. The fear of failure as well as the fears associated with success can often leave us experiencing emotional paralysis (fear) and can make moving forward in any direction difficult. In the same way, our lack of motivation to accomplish things in life (laziness) can set the stage for a lack of confidence (an unsound mind) that our efforts will be worth the exertion.

This circular loop, however, can create an inability to even recognize what is truly desired, which in turn, can lead to hopelessness. So what can we do to lift ourselves from the stagnation of perpetually following the paths of least resistance? Above all, we must embrace new thoughts and new words that align with the Word of God, which will raise us to new levels like a boulder is lifted up on a mountain. With an arsenal of new thoughts (as electrons) that counter those associated with the paths of least resistance, potential energy (as faith), can be stored and released at an appointed time. This potential energy was readily evident when the woman touched Jesus's clothes, or when it could be said that she resisted Not releasing the potential energy of faith. For God did not give her a spirit of fear but of power and of love and of a sound mind (see 2 Timothy 1:7).

# FEARLESS POTENTIAL

When the subject of potential comes up, what images come to mind? If you're anything like me, images of everything that I could be doing now but am not, usually show up. As a result, one of two responses occur: Either I quickly bury all thoughts regarding my own laziness and/or fear, or I quickly set my mind toward the work to be done. It turns out that work is the antidote to unfulfilled potential, and the destroyer of potential is laziness and fear. We understand this as we look back on those times where we have taken the paths of least resistance: Things of value are rarely produced during such times. So if taking the paths of least resistance keeps us from fulfilling our potential, the force of resisting is obviously of paramount importance in fulfilling one's potential.

Understanding the concept of potential can best be described in terms of energy. If we were to hoist a five-hundred-pound boulder to the top of a cliff, the same amount of energy that it took to get the boulder up to the top of the cliff (kinetic energy) would conceptually be stored in the rock as potential energy. Any expenditure of energy over time and space is considered work, which is expressed as (force x distance), and the potential energy of an object depends upon its position. Therefore, after being hoisted to the top of a cliff, an equal amount of stored or in this case gravitational potential energy as the energy it required to get it up on top of the cliff would remain latent in the rock indefinitely. Should the rock be pushed over the cliff, this potential energy would then suddenly be transformed into kinetic energy as it plummeted to the ground and destroyed anything in its way.

## LOGICS OF THE KINGDOM DYNAMICS

We can readily extend this understanding of physical potential energy when we think about athletes pushing themselves beyond impossible levels. Through resistance of physical pain or discomfort, new heights of achievement are met. But potential energy exists within all of us and requires the same measures to be fulfilled and is vulnerable to the same enemies to be quenched: Work must be done and effort must be expended for potential to be stored. Then that potential can have the greatest impact when it is jostled from its resting place and let loose in an appropriate environment; it's only when we have stored potential within us that we can go the distance with the right force inside of us to accomplish our dreams as well as God's will for our lives.

But let's not ignore the enemies of potential: Laziness and fear. While the root of one is the lure of immediate gratification, the root of the other is unbelief in the promises of God, which are favorable images and good reports. Both enemies have one thing in common: They lack vision, which exists for all as a goal point, like the summit of a cliff, and "Where there is no vision, the people perish" (Proverbs 29:18). Both fear and laziness keep us from moving forward—behaving as a strait jacket of sorts—and have as their only goal the limitation of our ability to experience the promises of God, as well as our ultimate destruction. Work, however, involves a force projected forward over a specific distance. So where there is work, there has been the hope of fulfilling a vision or goal.

While work may be the antidote to fear and laziness, when speaking of potential in terms of God's kingdom, work is best described in John 6:28-29 (NKJV) when Jesus answered the question, "What shall we do, that we may work the works of God?" In response, "Jesus answered and said unto them, 'This is the work of God, that ye believe on Him (Jesus) whom He (God, the Father) sent.'" So according to the kingdom of God, work involves faith. And as the scientific formula of work is work = force x distance, we need only look at Hebrews 6:10-12 for the other element of the equation:

> For God is not unrighteous to forget your work and labour of love, which ye have shewed toward his name, in that ye have ministered to the saints, and do minister. And we desire that every one of you do shew the same diligence to the full assurance of hope unto the end: That ye be not slothful, but followers of them who through faith and patience inherit the promises.

So whatever enables the followers of God to obtain the promises of God (which surpasses any potential that man by himself may have), requires faith and patience. And this is the work that we have been called to do, above all, which provides us the opportunity to enter His rest, having fulfilled the potential within us.

# A RIGHTEOUS THEOREM

Years ago, when I was trying to systematically uncover truths within mathematics, I was meditating on the Pythagorean Theorem, where the sides that make up the right triangle are A and B and the connecting line is C, or the hypotenuse. I had already uncovered an interesting correlation between a right triangle and righteousness. Not only do they have "right" as part of their grammatical construction, but righteousness alludes to a transparency or lack of deception. One fact that always struck me as significant was that when light passes through pure water, there is no refraction, meaning that the light will not bend in another direction, but it will remain traveling at a ninety-degree angle.

This got me to thinking about right triangles, and I began to see all three sides of the triangle as symbolic of the Christian walk. The ninety-degree vertical line represents our connection to God, but the horizontal line represents our connection with the world. And the hypotenuse is a graphic measurement of where we are in our walk with the Lord. If we have a short vertical line and a very long horizontal one, when the hypotenuse connects the ends it will describe a "walk" where the world is of greater importance. However, if our vertical line is tall and our horizontal line is short, the hypotenuse will describe a "walk" that is more determinately concentrated on the things of God. One is wide and low to the ground, having the force of gravity and all things more strongly associated with the grave in control. The other is tall and reaching new heights with just enough width at the base to maintain balance.

# WAVEFUNCTION COLLAPSE

Having bridged the Word of God with the four fundamental forces as well as with various other scientific processes, a while back I sought to bridge the Word of God with quantum mechanics (QM). During my research and deliberations, a revelation struck my heart: Since particles in QM are not measured according to their position or momentum, due to these measurements' mutual exclusion (Heisenberg principle), but are instead provided a wavefunction, which fundamentally describes the domain upon which they may be found, QM is all about probability and not certainty. Though obvious to many, this flashed like lightning within me—I realized that while it certainly describes several powers that are in the earth and under the earth, it was likely that I would not find scriptural reference to its application to the system in the kingdom of God. Why?

God is not a God of probability but is a God whose ways are certain, and His ways always adhere to His Word. He wants those in alignment with His Word to operate in the certainty of faith.

> Being fully persuaded that what he had promised, he was able also to perform (Romans 4:21).

In 1 Corinthians 9:26-27, Paul confirms this:

> I therefore so run, not as uncertainly; so fight I, not as one that beateth the air: but I keep under my body, and bring it in to subjection: lest that by any means, when I have preached to others, I myself should be a castaway.

## LOGICS OF THE KINGDOM DYNAMICS

Paul related a fundamental truth about God: There is no uncertainty in God, as He does what He says He will do, eventually, at least. The problem comes with those of us who fail to take Him at His word.

These were my thoughts regarding the concept of wavefunction, and quantum mechanics in general. I had all but given up trying to bridge it with God's kingdom principles until the concept of wavefunction collapse presented itself. Needing clarification about how the concept of wavefunction collapse related to God's heavenly system, I took a walk, which is what I often do to gain clarity; there is something about motion amidst the trees.

I understood the premise of a wavefunction, which was the notion that the probability distribution of finding a particle in a specific place assumed a wave-like graphical shape. Just like the wave-like shape of a roller coaster track travels in space, and as a roller-coaster car travels on the track at high speeds, so, too, is this wavefunction: Any attempt to locate the car on the track at any given time would almost be impossible to ascertain outside of relying on probability. As with the roller-coaster car, the probability of finding a particle in a specific place would remain in a probabilistic domain, where no certainty as to its location exists. That was basically my understanding of wavefunction, but in terms of wavefunction collapse, I couldn't quite get a handle on what this collapse might look like—until that walk.

After meditating upon this material, the Lord showed me a vision in a flash. It was a picture puzzle with one hundred pieces, all fitting neatly together with no missing pieces. He showed me that this was representative of any vision His children obtain by faith, having meditated upon His Word, because "Faith comes by hearing, and hearing by the word of God" (Romans 10:17 NKJV). When we have a vision for something that God has promised in His Word, this vision becomes like a picture puzzle in its entirety with no missing pieces. This is a vision that Habakkuk 2:2-4 speaks about as needing to be written and made plain upon tables so those who have read it may run after the vision:

> Write the vision, and make it plain upon tables, that he may run that readeth it. For the vision is yet for an appointed time, but at the end it shall speak, and not lie: though it tarry, wait for it; because it will surely come, it will not tarry.

While the whole picture puzzle represented a vision of things desired and hoped for, the concept of wavefunction collapse became clear: It would be as if the whole puzzle picture collapsed into a singular puzzle piece. When hope transitions into faith, for "Faith is the substance of things hope for" (Hebrews 11:1), the vision is certain to become a reality or substantial. However, what often happens is that people's faith fails and the visions "collapse." Circumstances and trials shift their focus from the promises of God to the natural facts playing out around them, and this shrinkage of a faith-filled vision into the individual cares and concerns of this world describes the concept of wavefunction collapse most accurately.

When we examine the notion of a faith-filled vision (picture puzzle) versus a singular puzzle piece, we get a glimpse of the difference between truth and fact. Just as picture puzzles are made up of singular puzzle pieces, the truth is composed of many facts. But because we will never arrive at the truth when we focus on a singular fact, visions cannot transpire when our focus is on the details of circumstances. And here lies the significance of wavefunction collapse—when our visions shrink in response to facts, the truth of God's promises can no longer describe a certain and promising reality.

# MAY I HAVE THIS...
## INTERFERENCE?

Human relationships can be a lot like dancing: leaders seek followers and followers seek leaders, but attention is often focused on themselves. In the case of ballroom dancing, however, attention must be on the complementarity of movements, or on the effects of one's stride on the other. And when partners are in step with one another, a beauty and gracefulness is sure to ensue, but when they are out of step? Oh boy! The effects range from an awkward stagger to the painful squashing of toes.

We've all experienced people with whom we just naturally flow: the conversation just happens fluidly and there is an overall sense of well-being that just takes over. We've also experienced people with whom everything just misses: the conversation is like pulling teeth and when your eyes meet, there is an automatic pulling away. Is it just us? Or is it just them? What's going on?

As dancers respond to music and the rhythms and melodies that emit from the instruments played, we also respond to frequencies that those with whom we "dance" or communicate emit. As Matthew 6:22 says:

> The light of the body is the eye: if therefore thine eye be single, thy whole body shall be full of light. But if thine eye be evil, thy whole body shall be full of darkness. If therefore the light that is in thee be darkness, how great is that darkness!

## LOGICS OF THE KINGDOM DYNAMICS

I interject this scripture not to bring judgment, but to open our eyes to a dynamic that is playing out. The way in which light travels is key! Light travels in waves, as does sound; therefore, when waves interact, they *interfere* with one another. And they can either constructively interfere or destructively interfere with one another.

When there is an attitude of excitement and emotions are high, the amplitude of the emitted wave could be regarded as positive. When a positive amplitude meets another person with a positive amplitude that happens to be traveling on the same frequency—where crest meets crest and trough meets trough—there is constructive interference. But when the trough of someone's wave meets the crest of another's, their waves destructively interfere. The result of this is often a flat line, or the end of conversation.

From this we can see why relationships can be a challenge. Like dancing, it takes patience, practice, and being in step, or "in phase," with one's partner. When our light goes forth from our eyes or when we speak words, we are emitting waves that collide, and this is where communication begins. There is also a coupling of light and sound as is revealed in the dynamic of thunder and lightning. The sound of thunder traveling more slowly than light is a response to a lightning bolt's collapse or contraction. This explains why certain words and sounds can register on the scale of good to bad and can begin the process of peace or war.

When we are interacting with our fellow human beings, therefore, the results are all about how our light and sound waves interfere. But remember, as with dancing, it always takes two to tango! Our waves' emittance can calm things down or ramp things up! If we want peaceful resolutions, it is within our power to seek peace in the Word of God. As Jesus said:

> Peace I leave with you, My peace I give to you; not as the world gives do I give to you. Let not your heart be troubled, neither let it be afraid (John 14:27).

John 8:31-32 (NKJV) also says:

> If you abide in My word, you are My disciples indeed. And you shall know the truth and the truth shall make you free.

In all relationships, we want to feel free, and we can affect change in the dynamics if we look, listen, and adjust our own response. For instance, in order to promote harmony, one needs to, above all, resist escalating chaotic or turbulent amplitudes by invoking a posture of love and/or understanding.

# REFLECTIONS

T hanks to Aristotle's logic, we know that since "white is reflective, and snow is white, snow is reflective." In the same way, we can understand that since God is love (see 1 John 4:8) and the Word was God (see John 1:1), the Word operates according to love. In addition, the Word was the "light of men" (John 1:4), and John 1:14 (NKJV) declares that "The Word became flesh and dwelt among us, and we beheld His glory, the glory as of the only begotten of the Father, full of grace and truth." From this we know that Jesus was the Word, He was the Light, and He was the same as God the Father. Therefore, we may conclude that God, Jesus, the Word, Love and Light are interchangeable, with reflection interestingly being one of the greatest common traits.

What is reflection? Aside from being a serious thought or consideration, it refers to an image seen in a mirror, or the throwing back of light, heat, or sound from a body or surface. As God and Jesus are considered Light, Matthew 6:22-23 provides an interesting distinction between differing lights:

> The light of the body is the eye: if therefore thine eye be single, thy whole body shall be full of light. But if thine eye be evil, thy whole body shall be full of darkness. If therefore the light that is in thee be darkness, how great is that darkness!

From this we understand that light can either be Light light or dark light; while Light light reflects spectral frequencies, dark light appears to absorb them, swallowing them up like a black hole. In our relationships, therefore, light is either exchanged fluidly with peace, joy, and love

## LOGICS OF THE KINGDOM DYNAMICS

flowing freely, or it is quenched and snuffed out with either a lack of care or the governing emotions of fear and/or hatred.

Reflective light emulates the love of God, which was alluded to when Jesus spoke to His disciples and said:

> Ye are the light (love) of the world. A city that is set on a hill cannot be hid. Neither do men light a candle, and put it under a bushel, but on a candlestick; and it gives light (love) unto all that are in the house (Matthew 5:14-15).

And He gave them a command, "Let your light (love) so shine before men, that they may see your good works, and glorify your Father which is in heaven" (Matthew 5:16). By this, Jesus was encouraging His disciples to love others as God Himself loved the world. A good example of this reflective light of love is seen in First Corinthians 13:4-8 (NKJV):

> Love suffers long and is kind, love does not envy, love does not parade itself, is not puffed up; does not behave rudely, does not seek its own, is not provoked, thinks no evil; does not rejoice in iniquity, but rejoices in the truth; bears all things, believes all things, hopes all things, endures all things. Love never fails.

This describes the reflective light of love, and as it is the Word, it also describes Father God and His Son, Jesus. Getting back to light, though, as everyone has been "fearfully and wonderfully made" (Psalm 139:14), each one of us is unique and has been endowed with specific gifts, and we emit different lights. We are no different than every other element on earth, and as every element is known by their spectral emission, we are no different.

In response to absorption of an incoming photon's energy, an electron in an atom absorbs the light, jumps to a higher level, and then releases the excess energy in the form of light. This process is exactly what occurs as people accept the Word and love of God into their heart.

The energy of the Word and love makes our spirits jump to a higher level, and as we return to an equilibrium state, we, too, release light energy through our eyes.

This light emission then becomes an identifier of who and what we are, and may, in fact, be how God recognizes each of us individually. As the spectrum is made up of every hue of every color, and this spectrum makes up white light, our positional relationship to the body of Christ and God Himself comes into greater focus. In the body of Christ, each individual believer supplies a light frequency so when put together with the rest of the body, the result will be the white light or love of God and His Son, Jesus.

Ephesians 4:15-16 (NKJV) alludes to this relationship:

> Speaking the truth in love, (we) may grow up in all things into Him who is the head—Christ— from whom the whole body, joined and knit together by what every joint supplies, according to the effective working by which every part does its share, causes growth of the body for the edifying of itself in love.

First John 4:17 declares that "As He is, so are we in this world." Like a picture puzzle, the body of Christ is made up of all different pieces (or light frequencies) that cannot illustrate the fullness of Jesus until every piece (or light frequency) comes together.

While this may be an illustration of perfection, as individuals we can reflect the love of God alone and, like God, this reflection is free of judgment, as Jesus commanded: "Judge not, that you be not judged." God stands before people as love and white light and reflects what is in their own hearts; He is like a white reflective surface against which their true nature will appear. And because He only reflects to us our own true nature, we do not feel condemned, but rather confronted with the truth. Now, the truth can hurt; it is true, but this provides an opportunity for repentance.

## LOGICS OF THE KINGDOM DYNAMICS

God's grace in action is revealed in Romans 2:4: "The goodness of God leadeth thee to repentance." When we repent, we turn from the darkness of this world and follow the reflecting light of the Word:

> For the word of God is quick, and powerful, and sharper than any twoedged sword, piercing even to the dividing asunder of soul and spirit, and of the joints and marrow, and is a discerner of the thoughts and intents of the heart. Neither is there any creature that is not manifest in His (the Word's) sight: but all things are naked and opened unto the eyes of Him with whom we have to do (Hebrews 4:12-13).

When our own light is Light light, which reflects and listens to others and shares the love of God with them, we verify 1 John 4:17, which declares, "As He is, so are we in this world."

# STOP! I WANT TO GET OFF!

D o you ever find yourself with a thought swirling around in your head repeatedly with no apparent plan to leave? It sticks around like a bad actor rehearsing a scene you would rather forget! So often these thoughts are like an annoying song getting stuck on repeat. Like water torture, they drip metronomically to a steady beat, until we are left screaming inside. Sound familiar? Instead of resigning ourselves to be a victim of these thoughts, let us realize, once and for all, that we are in the driver's seat!

After all, thoughts are simply light impulses that travel on paths within our brains, and they can be understood as functionally equivalent to electrons. Aside from governing many relationships between atoms, electrons, like thoughts, can travel in streams, and according to Andre-Marie Ampere's circuital law, discovered in 1826, aside from gaining momentum (and thus making it more difficult to stop), electric currents can travel in loops. And it is this dynamic of circular travel in loops that we have the physics behind what happens when a stream of thought seems to take over our brains, going around and around like the wheels on a bus.

An interesting thing to note is that when thoughts travel in loops, our emotions are stirred; the more we think about an injustice done to us, the more upset we become; or the more we think about an upcoming vacation, the more excited we are likely to be. Just like the electric and magnetic fields are always positioned perpendicularly to one another and a change in one causes a corresponding change in the other, so, too, are the fields that govern our thoughts (electrical impulses) and

## LOGICS OF THE KINGDOM DYNAMICS

emotions (the magnetic dynamics of those electrical impulses). In fact, the electromagnetic field itself can be understood as governing this thought/emotion field.

Just like a change in the electric field generates a magnetic field, we can relate this to a change in our thoughts generating a change in our emotions. A change in the emotional realm also always affects the realm of thoughts. If an old friend suddenly sneaked up behind you and gave you a friendly embrace, before you had time to generate any thoughts, your emotions would be positively affected, and thoughts of well-being and hope would spring forth.

So what can we do when intruding thoughts are taking over our minds as well as our emotions? Begin to *say* the opposite of what you have been thinking. Read. Sing. Speak out loud the things that you desire! This will disrupt the thought pattern and will introduce a new thought that can have an immediate effect on your emotions: "For as (a "man") thinketh in his heart, so is he" (Proverbs 23:7). Our state of being, to a large extent, is contingent on our emotions, and with our thoughts governing our emotions, this relationship between thoughts and emotions is once again confirmed. There is also a dynamic called the "Saying is believing paradigm" which contends that the more we say something, the more our minds and our hearts will be convinced that it is true.

This provides a clue into how to disrupt a negative train of thought, and by introducing elements that affect the five senses, we can likewise change our emotions and feelings. Instead of being brought to a place where our minds are screaming, "STOP! I WANT TO GET OFF!" we can rather reside in a place where we are in control of our thoughts and emotions instead of having *them* in control of us.

# THE GRAPHITE PRINCIPLE

When we think about precious stones, a diamond is one of the first to come to mind. While the mineral's uses now extend into the world of electronics as a semiconductor, it is most notable as the premiere choice for wedding rings, due probably to the delightful ways in which the light bounces off the chiseled planes. Made up of carbon atoms with a strong covalent bonding that are arranged in a repeating pattern of eight atoms called a diamond lattice, this atomic structure seems to hold a key to its strength, which is a defining characterization.

Validating an importance of structure over substance, not only is eight the number of completion, as in an octet (an atomic shell with eight electrons which creates stability), but at the other end of the durability spectrum lies another purely carbon substance called graphite, with none of diamond's qualities. It's interesting because while these two natural objects are made up of the same exact element, due to the molecular structure, a diamond is one of the hardest substances and graphite, which is made up of carbon sheets, is one of the weakest.

A hidden reference to these materials may, in fact, exist in Matthew 21:42-44. In responding to the chief priests and elders of the people who had challenged Him regarding His teachings after providing a couple of parables, Jesus said to them:

> Did ye never read in the scriptures, The stone which the builders rejected, the same is become the head of the corner: this is the Lord's doing, and it is marvellous in our eyes? Therefore say I unto you, the kingdom of God

# LOGICS OF THE KINGDOM DYNAMICS

shall be taken from you, and given to a nation bringing forth the fruits thereof. And whosoever shall fall on this stone shall be broken: but on whomsoever it shall fall, it will grind him to powder.

In this scripture, we understand that the stone which the builders rejected is Jesus, who also is the Lamb to whom the body of Christ shall be wedded. But we must understand an implicit message in the Bible, which is that if anyone humbles themselves enough to come to God, he is exalted; but if they resist and deny Him, they may find themselves crushed.

While God is love (1 John 4:8), He is also just, rewarding according to our deeds, particularly when the grace of our Lord Jesus Christ has not been accepted by faith. Herein we may find these substances as those being alluded to in Matthew 21:42-44. If Jesus is this stone, and He is the firstborn of many brethren, we, too, are made up of the same "material," according to 1 John 4:17, which says, "As he is, so are we in this world." But notice that this stone is something that can be fallen upon in addition to something that can crush and destroy. Now if anything were to fall upon a diamond, wouldn't it be broken, as diamond is one of the hardest materials? And if we were like Jesus, but not built with the structure of a diamond lattice, we may be configured like the planar sheets of graphite, which would be crushed should a diamond fall upon it.

But not only do these substances differ in hardness, they also differ in function.

While a diamond has the highest hardness and thermal conductivity of any bulk material, graphite is considered a non-metal that conducts electricity. However, due to one of the electrons in each carbon atom not being delocalized, meaning that it is not free to conduct electricity in all directions, graphite's ability to conduct electricity is limited; it cannot conduct electricity upwards.

When I studied this, I instantly understood the significance of Matthew 21:44. Graphite's limitation of only conducting electricity sideways describes people whose relationships are only with other people. Whether they acknowledge God or not, there is no connection with Him. And when there is no connection with God, people will not *"fall"* upon Him, or rather collapse within His loving arms, as a Father whom one trusts. This of course takes humility, a faith that God exists and a desire to turn toward hope, which draws us toward Him.

This drawing forth describes God's magnetism, which is a great illustration of repentance and can be understood as making a complete change of direction. God's magnetic force is like a giant bar magnet whose "negatively-charged" side causes any object with a similar charge that draws near to it, to turn around. When the smaller "negatively-charged" body turns, its "positively-charged" side becomes magnetically bound to the larger magnet. And here we see Romans 2:4 play out, as it is the goodness (or grace) of God that leads one to repentance, and we get a picture of this magnetic connection as the ultimate expression of God's love.

When people fail to "fall" upon God with humility, they are setting themselves up to not only remove themselves from the love of God but also to be fallen upon. And Matthew 21:44 comes back to mind as well as the thought of graphite: "And whosoever falls on this stone shall be broken; but on whomsoever it shall fall, it will grind him to powder." When we don't seek to have an intimate relationship with God who can reside within us, and seek only relationships with other people, we are like the substance of graphite whose structure enables it to be crushed should it be fallen upon. But we can invite the "stone which the builders rejected" (Jesus) who "has become the chief cornerstone" into our heart (Mt. 21:42), and become like a diamond that can reflect the light and love of God back into the world.

*Where there is no vision, the people perish.*

Proverbs 29:18

# OUR PRO-VISION

While God is our provider, one thing is required of us: We are to hold fast not only to our confessions (Hebrews 4:14 NKJV) but to our visions as well, for "Where there is no vision, the people perish" (Proverbs 29:18). We are to be pro-vision minded as opposed to anti-vision minded, keeping the picture of what we want or have been promised before our mind's eyes.

After Moses died, Joshua was told by God that he was to lead the children of Israel into the promised land:

> Only be strong and very courageous, that you may observe to do according to all the law which Moses My servant commanded you; do not turn from it to the right hand or to the left, that you may prosper wherever you go. This Book of the Law shall not depart from your mouth, but you shall meditate in it day and night, that you may observe to do according to all that is written in it. For then you will make your way prosperous, and then you will have good success (Joshua 1:7-8 NKJV).

Whatever God has told us to do, we, too, are to be strong and very courageous and hold fast to the words of promise that declare our ability to obtain it. As it is impossible to please God without faith (Hebrews 11:6), it is impossible to reach our hopes, goals, and desires without maintaining the vision before our eyes and expecting the vision to come to pass.

# THE APPOINTED TIME

Declaring the importance of holding firm to prophetic words or visions, Habakkuk 2:2-3 proclaims:

> Write the vision, and make it plain upon tables, that he may run that readeth it. For the vision is yet for an appointed time, but at the end it shall speak, and not lie: though it tarry, wait for it; because it will surely come, it will not tarry.

If we believe this scripture, nothing should ever surprise us; every word that has been written in the Bible is a vision or prophecy that has, is coming, or will come to pass. Having said that, regarding the birth of John the Baptist, this event seems to have begun the times of all prophetic fulfillment:

> Among them that are born of women there hath not risen a greater than John the Baptist: notwithstanding he that is least in the kingdom of heaven is greater than he. And from the days of John the Baptist until now the kingdom of heaven suffereth violence, and the violent take it by force. For all the prophets and the law prophesied until John. And if ye will receive it, this is Elias, which was for to come (Matthew 11:11-14).

Before John, the law and the prophets prophesied, which means that they spoke of things that had yet to come to pass. So John marked the arrival of such things. For this reason, he is esteemed the greatest among those who are born among women. Not only was he a man who came

*LOGICS OF THE KINGDOM DYNAMICS*

forth from the womb of a woman through the process of birth, which, like nuclear fission, indicates a "break-through" event, but he was the anticipated prophet, Elijah, who would be sent before the coming of the great and dreadful day of the Lord, as mentioned in Malachi 4:5. In addition, he was also the prophetic fulfillment of Isaiah's prophecy:

> The voice of him that crieth in the wilderness, Prepare ye the way of the Lord, make straight in the desert a highway for our God. Every valley shall be exalted, and every mountain and hill shall be made low: and the crooked shall be made straight, and the rough places plain: and the glory of the Lord shall be revealed, and all flesh shall see it together: for the mouth of the Lord hath spoken it (Isaiah 40:3-5).

All prophetic utterances have an appointed time of fruition, and when the "seed" of a prophetic vision is planted, nourished, and watered with the Word of God, it will (in due time) split; and the sprouts of life regarding the prophecy will break through the soil and reveal themselves. This is what happened when John the Baptist entered the scene. He seemed to satisfy many prophetic utterances, including references to him as being the prophet Elias or Elijah as stated in Matthew 11:14 as well as Malachi 4:5-6:

> Behold, I will send you Elijah the prophet before the coming of the great and dreadful day of the Lord: and he shall turn the heart of the fathers to the children, and the heart of the children to their fathers, lest I come and smite the earth with a curse.

So what happened from the time of John the Baptist until the "now" that was spoken of in Matthew 11:12? When John's mother, Elisabeth, was six months pregnant, John was baptized by the Holy Spirit by meeting Mary, Elisabeth's cousin, who was carrying Jesus at the time. Not only was he the first man to be baptized in the Holy Spirit via Jesus, but he was the man who ushered in the prophetic fulfillment of the

Lord's Christ, the savior of the world. He broke the proverbial hymen of prophetic manifestation, with Jesus (the light) following, which is confirmed in Matthew 3:1-3 and Matthew 4:13-17:

> In those days came John the Baptist, preaching in the wilderness of Judaea, and saying, Repent ye: for the kingdom of heaven is at hand. For this is he that was spoken of by the prophet Esaias, saying, The voice of one crying in the wilderness, Prepare ye the way of the Lord, make his paths straight.

> And leaving Nazareth, he came and dwelt in Capernaum, which is upon the sea coast, in the borders of Zabulon and Nephthalim: that it might be fulfilled which was spoken by Esaias the prophet, saying, the land of Zabulon, and the land of Nephthalim, by the way of the sea, beyond Jordan, Galilee of the Gentiles; the people which sat in darkness saw great light; and to them which sat in the region and shadow of death light is sprung up. From that time Jesus began to preach, and to say, Repent: for the kingdom of heaven is at hand.

So what was going on "from the time of John the Baptist until now"? The preaching of the kingdom! Luke 16:16 appears to equate the preaching of the kingdom to the kingdom of God suffering violence, as mentioned in Matthew 11:12-13. Here again, we see that the advent of John is a marker, of sorts, for the end of the law and the prophets, as these scriptures state:

> The law and the prophets were until John: since that time the kingdom of God is preached, and every man presses into it. And it is easier for heaven and earth to pass, than one tittle of the law to fail (Luke 16:16-17).

This scripture indicates that John's appearance coincided with the fulfillment of the law and the prophets, and Matthew 5:17 confirms that

## LOGICS OF THE KINGDOM DYNAMICS

this was Jesus's purpose as well: "Think not that I am come to destroy the law, or the prophets: I am not come to destroy, but to fulfill."

The mere fact that Jesus had to proclaim that He was not there to destroy but to fulfill, indicates that something of a violent nature was occurring, at least in the spiritual realm. And there is no better illustration of this sort of violence occurring simultaneously to a fulfillment of something than the process of birth itself.

If the time of John the Baptist and Jesus indicates fulfillment of the law and the prophets, the law and the prophets indicate a calling forth of those things that be not as though they were (Romans 4:17). The response of violence to the manifestation of prophetic utterances is interesting, because we see this occurring right now. Donald Trump's presidency seems to be a shadow of things to come. I don't believe that it's any coincidence that 1 Corinthians 15:52 declares, "In a moment, in the twinkling of an eye, at the last *trump*: for the trumpet shall sound, and the dead shall be raised incorruptible, and we shall be changed."

If Donald Trump is a symbolic representation of *"the last trump,"* we are witnessing the violence alluded to as a prophetic fulfillment, with the nonsensical outbursts of violence and hatred demonstrated day and night against this United States' president. It is also interesting to note Isaiah 32:5, which says:

> The vile person shall be no more called liberal, nor the churl said to be bountiful. For the vile person will speak villany, and his heart will work iniquity, to practise hypocrisy, and to utter error against the Lord, to make empty the soul of the hungry, and he will cause the drink of the thirsty to fail. The instruments also of the churl are evil: he deviseth wicked devices to destroy the poor with lying words, even when the needy speaketh right. But the liberal deviseth liberal things; and by liberal things shall he stand.

These vile and churlish people are the violent who are attempting to "take" the prophetic fulfillments by force, trying to keep them from manifesting. However, God's Word stands sure:

> For as the rain comes down, and the snow from heaven, and do not return there, but water the earth, and make it bring forth and bud, that it may give seed to the sower and bread to the eater, so shall My word be that goes forth from My mouth; it shall not return to Me void, but it shall accomplish what I please, and it shall prosper in the thing for which I sent it (Isaiah 55:10-11).

The vile (or the violent), however, are simply flesh and blood against whom we do not war; they are pawns in a battle that has already been won. The true violence that the kingdom of heaven suffers is the war between the dragon and his angels and Michael and his, as recorded in Revelation 12:7-12:

> And there was war in heaven: Michael and his angels fought against the dragon; and the dragon fought and his angels, and prevailed not; neither was their place found any more in heaven. And the great dragon was cast out, that old serpent, called the Devil, and Satan, which deceiveth the whole world: he was cast out into the earth, and his angels were cast out with him. And I heard a loud voice saying in heaven, Now is come salvation, and strength, and the kingdom of our God, and the power of his Christ: for the accuser of our brethren is cast down, which accused them before our God day and night. And they overcame him by the blood of the Lamb, and by the word of their testimony; and they loved not their lives unto the death. Therefore rejoice, ye heavens, and ye that dwell in them. Woe to the inhabiters of the earth and of the sea! for the devil is come down unto you, having great wrath, because he knoweth that he hath but a short time.

## LOGICS OF THE KINGDOM DYNAMICS

And this is yet another vision or prophecy that *will* come to pass. "He (the devil) has but a short time!" The appointed time!

# DEVELOPABLE FAITH

While encouraging his followers in Rome to live their lives in such a way that everything they did was pleasing to God, Paul instructed them to:

> Be not conformed to this world: but be ye transformed by the renewing of your mind, that ye may prove what is that good, and acceptable, and perfect, will of God (Romans 12:2).

This was good advice, as it is abundantly clear just how much the world's ideas and perspectives endeavor to overtake all who might be overcome, and these ideas do, indeed, have an effect! Not only do they engender strife by attacking all thoughts that oppose them, but more destructively, they paint pictures of outcomes that serve death rather than life.

These pictures take root inside people's heads, and through fear (which is the inverse of faith), these images can be transformed into reality. This is what Paul was identifying in his speech to the Romans, for he—or at least the spirit within him—knew that if people were going to fulfill the vision that God had for their lives, the images in their heads MUST be transformed!

This transformation of thought is the cornerstone of what is declared throughout the Bible: "The just shall live by faith" (Habakkuk 2:4; Romans 1:17; Galatians 3:11; Hebrews 10:38). As fearful thoughts can generate negative outcomes for those who entertain them, so, too, can "faithful" thoughts that align with the blessings of God generate those

## LOGICS OF THE KINGDOM DYNAMICS

same blessings in the lives of those who entertain *them*. It truly all comes down to thoughts and pictures that are generated from a larger source.

When you think about the sources of information that generate pictures today: Television and cable news; movies; TV talk shows; Facebook; Twitter; newspapers; etc., much of these simply are disseminating (or sowing) thoughts that conform to this world into the minds of its viewers. All pictures are silently bound to words, and it is for this reason that visions speak and have power. And they are piggy-backing on a process that reveals how both faith as well as fear become foundations upon which we may live our lives.

Around a hundred and forty years ago, capturing real-life images—being able to generate a photograph—was the premier achievement in terms of being able to lock in a process that would forever change how people's minds were influenced; as the saying goes, "a picture says a thousand words." The problem is that we were created to be governed by the words that we speak, for "Death and life are in the power of the tongue" *(Proverbs 18:21)*. But with the inundation of images comes the inundation of generated thoughts, which often correspond with death rather than life. In other words, images have usurped our power to frame our world view and take our rightful dominion and authority over this world.

While we were created to govern our lives by the words we speak, we all live by the pictures in our own heads, and those pictures will lead us toward roads of success or roads of failure. The "why" behind this, however, has remained a mystery. People may understand the "Saying is believing" principle, which asserts that if people say something long enough they will begin to believe that it's true, as well as the placebo effect, where people's belief in the efficacy of a remedy will often be enough to gain a positive effect, even if the remedy is never actually taken. Both examples prove that there is a power in belief, even if it is not fully understood. But having lived by faith since I was nineteen years old, it has just recently become apparent just how imperative the transformation of thoughts truly is—and this vital

*HEATHER FORTINBERRY*

transformation can be understood in the contextual examination of the old photographic process.

Before the digital era, photographic images relied upon old-fashioned film, which is coated with gelatin emulsion in which silver halide crystals are embedded. Simply stated, when this film is exposed to light, a sensitivity speck on the surface of the crystal is turned into a small speck of metallic silver. How does this occur? When photons strike the silver halide crystals, de-localized electrons are generated, which migrate through the lattice structure until they are captured by a shallow electron trap, which is where the invisible or latent image resides.

The image of what was photographed exists in a hidden realm that requires light, electrons, and developing agents, which are then added and convert the latent image into a visible image by reducing silver halide crystals to silver atoms. This process, interestingly, mirrors the process of developing one's faith. And it all begins with eradicating the images that have overtaken our minds, as 2 Corinthians 10:5-6 states:

> Casting down imaginations, and every high thing that exalteth itself against the knowledge of God, and bringing into captivity every thought to the obedience of Christ; And having in a readiness to revenge all disobedience, when your obedience is fulfilled.

With the understanding that electrons and thoughts (both electrical impulses) are functionally equivalent, we can see the sequestration of electrons in the shallow electron trap as signifying the captivity of thoughts that one wants developed, having willfully rejected those that do not align with the will of God. And it is in this captivity where our thoughts can undergo fruitful conversion. As in an emulsion, where at least four atoms need to be within an electron trap for the entire image to be considered developable, in describing developable faith, we can identify specific steps in the process.

As in any photographic endeavor, the first step involves identifying what we want to photograph. What image has impressed us so much

## LOGICS OF THE KINGDOM DYNAMICS

that we want to keep it in our lives to behold and enjoy? This step also initiates any stand of faith; first, we must identify our desires. What can we see bearing fruit in our lives? The next step in the process is pressing the shutter button, which will release the aperture from its closed state. Once open, light will flood in and expose the film, generating free electrons.

The equivalent, spiritually speaking, is that once we have identified what we want, we must search the Word of God (light) for a promise. Then we must meditate on the promise until we are fully convinced that what God (the Word) has promised, He is able to perform (see Romans 4:21). *This* is the process that can be understood as that which converts the latent image of the subject of our faith into a visible image or reality.

When I was believing the Lord that I would become pregnant, having been told that I would not be able to conceive, I looked to Deuteronomy 28:4 (NKJV):"Blessed shall be the fruit of your body." For six months I meditated on and declared that scripture morning, noon, and night. I basically lived on that Word, according to Matthew 4:4 (NKJV): "Man shall not live by bread alone but by every Word that proceeds out of the mouth of God." In addition, I called those things that were not as though they were (see Romans 4:17), by saying "I *am* pregnant!" These were my constant declarations until the "Word became flesh and dwelt among us" (John 1:14 NKJV), which occurred on August 22, 1996.

# WHERE THERE'S VISION, THERE'S LIFE!

I have learned something over the years: If I cannot envision something happening, it will *NEVER* become a reality for me. If I cannot see myself performing some action, or just being myself without compromising who I am, I know that the situation will not bear fruit. I remember one day while I was sitting in a class at graduate school. I was thoroughly engaged in learning about systems and how they relate to human behavior, when an image suddenly flashed in my mind and I saw myself teaching the class, and I said to myself, "I will teach!"

While I had taught young children and had a degree in art education, I saw myself teaching the concepts that were fundamentally changing my life. Knowledge and understanding about why I and my family as well as others behaved the way we did, was of such an earth-shaking magnitude that the excitement in my heart could hardly be controlled. The vision caused a burst of light in my soul, and I knew that the excitement could bear fruit.

A year later I received a call from the film school where my ex-husband had taught. The woman on the other end had a question, "Would you be interested in teaching psychology?" she asked. "Yes!" I burst out. And the vision that I had seen became a reality. The vision of me teaching and the words, *I will teach,* became a declaration. Job 22:28 says:

> You will also declare a thing, and it will be established for you; so light will shine on your ways.

## LOGICS OF THE KINGDOM DYNAMICS

The vision came; I declared words to support the vision, and I fought the fears that told me I could never succeed. I had never been able to speak in front of people without stumbling over my words and becoming flushed and forgetting what I wanted to say. But I searched the Word for a promise, and "I can do all things through Christ who strengthens me" (Philippians 4:13 NKJV) became my mantra.

To go forward in life, goals (or pictures) of where we want to see ourselves in the future are essential, for "Where there is no vision, the people perish" (Proverbs 29:18). However, people live according to stories in their head, and sometimes those stories are not of success, but rather of failure. People tend to retain memories of failures more than those of success, as well as words of judgment more than words of praise. Searching our memories for exceptions to our stories, however, can provide the courage to push forward.

When I was faced with only memories of failure when being in front of people, I searched and searched until I remembered dancing on stage with my sister when I was six years old. And while it had not required me to speak, I had one vision of me succeeding in front of people on stage. That became a source of strength—knowing that I had it within me not to crumble. Habakkuk 2:2-3 declares a truth:

> Write the vision, and make it plain upon tables, that he may run that readeth it. For the vision is yet for an appointed time, but at the end it shall speak, and not lie: though it tarry, wait for it; because it will surely come, it will not tarry.

The vision of me teaching and not crumbling came, and established a firm understanding of how dreams and goals can bear fruit.

# CONCEIVE — BELIEVE — RECEIVE!

I learned a long time ago to expect those things that I speak to come to pass. After years of speaking words that might get a rise out of friends or might provide them with a laugh, I made the connection. Contrary to popular belief, there is an undeniable correlation between what one says and what one experiences, at least eventually.

I've had both the pleasure as well as the misfortune to confirm this phenomenon through experience; I have called "those things which be not as though they were" (Romans 4:17) in faith and have been doggedly determined not to speak words that were contrary, and I have received those things. I have also spoken things in fear and believed that my situation would not fare well, and I have received these things as well. While circumstances have not always looked exactly like the initial vision (whether of fear or of faith), upon deeper inspection, my life has been defined by the words that I have spoken. And while every word can be understood as a seed, the initial substance of that seed has always been a thought. Our thoughts are, therefore, vitally connected to our reality, for what we allow in our minds will come out of our mouths.

This is an integral part of the system of the kingdom of heaven. In the first chapter of Genesis, we see everything that was created was first spoken into existence. As "the Spirit of God moved upon the face of the waters" (Genesis 1:2) before God said, "Let there be light," we implicitly understand that this movement as well as these words happened in response to a thought, or rather, a conception. From this conception,

## LOGICS OF THE KINGDOM DYNAMICS

words were formed, which behaved as seeds planted in fruitful ground. Interestingly, the entire first chapter of Genesis appears to speak only of God's conception of the world. This is substantiated beginning in the fourth verse of Genesis chapter two:

> These are the generations of the heavens and of the earth when they were created, in the day that the Lord God made the earth and the heavens, and every plant of the field *before* it grew: for the Lord God had not caused it to rain upon the earth, *and there was not a man to till the ground* (Genesis 2:4-5).

But wait just one minute! Didn't Genesis chapter one declare that everything—planets, stars, plants, animals, as well as man—had already been made? But here we see that at least the plants and man had not actually grown or been formed until a mist went up from the earth: and so it seems that there is a distinction between "created" and "made." When God said: "Let us make"; "let there be"; "let the earth bring forth", etc., He was apparently involved in a "first creation" stage of creativity, whereby the conception of something new always begins in the mind, at the thought level. Genesis chapter one, therefore, is describing things that were primarily *created* or conceived by the mind of God.

As it has been written in Genesis 1:26 that God said, "Let us make man in our image, after our likeness: and let them have dominion," this was a conception of "man," who distinctly differed from other creatures in that man's identity was connected intimately with God's. And this connection imparted a greater consciousness to "man," which would allow them to have dominion over all the earth, for God could trust a people with His heart to dominate. The problem was that this class of mankind mingled his seed with the daughters of men (Genesis 6:2), and therefore God's plan for people with His heart dominating the earth didn't work out, that is until Jesus died so that those who believed might become sons of God once again.

*HEATHER FORTINBERRY*

The point that "man" was made with a creative power like God's cannot be undervalued. As Romans 4:17 contends, one of the main characteristics of God is that He calls "those things which be not as though they were." God brings forth with His words those things that are tucked away in the invisible realm. And we are no different, for good or for bad. When we speak with positive words about things that we have envisioned, we can accomplish anything we can imagine. However, as we live in a polarized world with opposite forces vying for our attention, we also can open the door to our fears manifesting as well.

As I said before, I have experienced both my hopes as well as my fears becoming a reality when I have spoken them into being, intentionally and unintentionally; this is real! You probably know at least one person whose words come to mind as having prophesied some end. Many people end up prophesying about their final hour to the horror of those who know them. Words brings much to pass, but do not neglect the power of thoughts as well, for they have every bit as much power to affect an eventual end. That's why Philippians 4:8 says what it does:

> Finally, brethren, whatsoever things are true, whatsoever things are honest, whatsoever things are just, whatsoever things are pure, whatsoever things are lovely, whatsoever things are of good report; if there be any virtue, and if there be any praise, think on these things.

Our thoughts, after all, govern our motivations.

While my own life is replete with examples, one example in particular comes to mind. I was born with a yearning to write, and I wrote poetry whenever I could. At the age of seven I began to label myself as a writer; that is what I liked to do, and therefore that is what I was. However, as the years passed, I rarely wrote anything and I was totally not confident in my English classes, except when it came to writing poetry, but still I continued to call myself a writer. After high school I was in a car accident, which resulted in me having a brain injury—and any thought of pursuing writing as a career seemed no longer viable, due to short-

## LOGICS OF THE KINGDOM DYNAMICS

term memory loss. My fears of failure were in the driver's seat at this point, and they prevented me from challenging myself.

I continued through those years, though, feeling a sensation akin to a "ghost limb" where the dream of writing had always been. In response, however, I began to call myself a writer again, calling those things that be not as though they were, just as Romans 4:17 declares. At the same time, I pursued a masters' degree in psychology, which required me to write papers, but nothing like what I imagined writing in my dreams; I had always yearned of being a fiction writer with a philosophical edge. But in 2006 when my son was in the fourth grade, I challenged myself and began to collaboratively write a story with his class, based on their ideas, story line, and characters. I self-published the book and gave every child his or her own copy with hopes that it would inspire them to read as well as to write. The writing is abysmal, but it was the challenge and experience and a prophetic fulfillment of what I had been saying for years: finally, I *was* a writer.

Two years later I found myself undertaking a project that has become my life's work, and I was able to write a 250-page book detailing a theory of the universe with God's Word at its center, with the spokes of the wheel touching scientific, spiritual, as well as psychological phenomena. Without those years of proclaiming that I was a writer, I don't believe the vision of me one day writing would ever have materialized:

> Therefore I say to you, whatever things you ask when you pray, believe that you receive them, and you will have them (Mark 11:24 NKJV).

Our thoughts and words draw an image in our minds of what we desire as well as what we fear. I encourage you to only think on those things that you want, so that when you speak, your words will render good and prosperous images. For what you conceive in your mind and believe in your heart, you need to speak forth out of your own mouth so your inner ear will hear the words; you will receive if you only believe!

# EVIDENCE

One of my normal practices is to walk my dog, Willie, and contemplate and meditate upon the Word while I walk, with quiet assurance that I will receive revelation. Years ago, I read Jeremiah 33:3 (NKJV), which says, "Call to Me, and I will answer you, and show you great and mighty things, which you do not know." I never cease to ask, and He has always been faithful! But as His thoughts are not our thoughts and His ways are not our ways (see Isaiah 55:8), His timing is not always the same as our timing, but He always opens and closes the right doors when we trust Him. Anyway, I was contemplating His promises both in the Bible as well as in prophetic utterances, and I began to meditate upon Hebrews 11:1:

> Now faith is the substance of things hoped for, the evidence of things not seen.

After a period of meditating on this scripture, the revelation of it came in an instant! Faith is the material *evidence* of all we need and desire from God! Faith is not just belief, as some might contend; but it is like the document in one's hands that proves one's innocence! It is an alibi! It is the original lottery ticket! And just as we hold dearly to these, we should hold dearly to our faith in what God has said in His Word or to us personally. When His words are securely held in our heart and in our mouth, they can overcome any other presenting fact. That is why Jesus said to His disciples, "Have faith in God" (Mark 11:22) in response to Peter asking about the fig tree drying up at Jesus' command. When we have faith that what the Lord has spoken is absolute truth and is sure to come to pass, we will have the promise in our possession!

# LOGICS OF THE KINGDOM DYNAMICS

> For assuredly, I say to you, whoever says to this mountain, "Be removed and be cast into the sea," and does not doubt in his heart, but believes that those things he says will be done, he will have whatever he says. Therefore I say unto you, What things soever ye desire, when ye pray, believe that ye receive them, and ye shall have them (Mark 11:23).

If we hold fast to our confession of faith and "see in our mind's eye" the result of belief, our faith (the steadfast vision held without wavering) is *all* the evidence we need to expect whatever we have asked for in prayer to come to pass!

> And this is the confidence that we have in him, that, if we ask any thing according to his will (His Word), he heareth us: And if we know that he hear us, whatsoever we ask, we know that we have the petitions that we desired of him (1 John 5:15).

As kings have been afforded the power to establish rules on earth, according to Job 22:28, we also can "Decree a thing, and it shall be established." Just as a king has faith that his mandates and decrees (His will) will be followed, so, too, can our faith in our king's (Jesus's) words bring about whatever is declared (God's will) and believed to be unequivocally true. Our faith that what God has said is truth and will come to pass at the set or appointed time, is the *proof* that we can furnish at the time of trouble, test, or trial. It is all we need to be able to stand in front of judges, juries, and enemies alike—we can confidently wave our evidence before them.

We can carry this proof with us, assured that we can stand against any enemy that comes against us. If we *receive* what has been provided to us, we have this proof in the form of a *receipt*. Carry that receipt with you everywhere you go, in your heart and in your mind, so shall you "flourish in the courts of our God" (Psalm 92:13). Our faith is all the evidence we will ever need! Guard that evidence with all your heart and with all your mind!

*I am Alpha and Omega, the beginning and the ending.*

Revelation 1:8

# BECOMING FRUITFUL

One of the first things that occurred in the very beginning was that God said:

> "Let the earth bring forth grass, the herb that yields seed, and the fruit tree that yields fruit according to its kind, whose seed is in itself, on the earth"; and it was so. And the earth brought forth grass, the herb that yields seed according to its kind, and the tree that yields fruit, whose seed is in itself according to its kind. And God saw that it was good (Genesis 1:11-12 NKJV).

That fruit yields seed, therefore, is among those things that are good, so when it's planted in good soil, that seed has the potential to bear good fruit. Bearing good fruit, however, is contingent on being a good tree, as Matthew 7:17-18 declares: "Every good tree bears good fruit, but a bad tree bears bad fruit. A good tree cannot bear bad fruit, nor *can* a bad tree bear good fruit." As we have been created to "Be fruitful, and multiply" (Genesis 1:28), Colossians 1:9-10 (NKJV) specifies that we are to "walk worthy of the Lord, fully pleasing Him, being fruitful in every good work and increasing in the knowledge of God." Therefore, we are called to be good trees.

We see that good trees bearing good fruit (and bad trees bearing bad fruit) follows the rule of self-similarity. All things replicate according to their kind, so what is planted is of vital significance. According to Luke 8:11 in the parable of the sower, "The seed is the Word of God." And Hebrews 11:3 declares that "Through faith we understand that the worlds were framed by the word of God."

## *LOGICS OF THE KINGDOM DYNAMICS*

It would appear that the worlds—as well as everything in them—were framed by a seed, which, (like all seeds) stores within itself a specific DNA sequence. DNA, therefore, accomplishes in the natural realm what the Word of God can accomplish in the spiritual realm with evidential results. When we sow the Word of God by declaring what God has already said, i.e., the promises that He has left us, we can be assured that that Word will bear fruit that is according to what has been spoken. For instance, "The fruit of my body is blessed" will eventually bear the fruit of having a healthy baby, should the words be mixed with faith, which is "the substance of things hoped for, the evidence of things not seen" (Hebrews 11:1).

When sown in well-fertilized soil, or rather a hearing heart, the Word of God has within itself the pattern of whatever is spoken, as the Lord spoke to my heart saying, "The Word is the DNA pattern of My will." When we speak according to God's Word and are "fully convinced that what He has promised He is also able to perform" (Romans 4:21 NKJV), we are planting a seed that will bring a harvest to fruition. Mark 11:23-24 clarifies:

> For verily I say unto you, That whosoever shall say unto this mountain, Be thou removed, and be thou cast into the sea; and shall not doubt in his heart, but shall believe that those things which he saith shall come to pass; he shall have whatsoever he saith. Therefore I say unto you, What things soever ye desire, when ye pray, believe that ye receive them, and ye shall have them.

As we know, words are essentially codes that stand for things in our world. DNA is no different, but it is composed of codons, which are groupings of three out of four chemicals. Arranged in different ways along chains, these codons describe every living thing, and through an involved process of being transcribed from a DNA template strand into a Messenger RNA strand and then translated into a protein, all the necessary information that can keep a species and/or individual organism alive is available. Just like the end-product of messenger RNA (protein) is sent out into the body as nourishment, so, too, is the Word of God sent out into the body of Christ. As life is given to every species by this DNA code, it is also by the Word of God that mankind can sustain life, as Matthew 4:4 declares:

> It is written, Man shall not live by bread alone, but by every word that proceedeth out of the mouth of God.

# FOOTPRINTS

For years now, I have illustrated a conceptual framework of the relationship between heaven, earth, and Christians as they walk in alignment with God's Holy Spirit and His Word. This illustration originally was just two horizontal parallel lines connected with a perpendicular line in the middle; but it has graduated to many perpendicular lines (delineating a ladder on its side) connecting the parallel ones. This ladder has represented heaven and earth as they are connected by the children of God on their life's journey, with each perpendicular rung representing a perfect adherence to the will of God, which is represented in the heavenly realm. This, of course, represents the ideal.

When God's children walk in the Spirit and not in the flesh (Romans 8:1), this ideal is realized. This has been a fitting illustration of God's will being done on earth as it is in heaven, just as the Lord's prayer, as written in Matthew 6:10 suggests, "Thy kingdom come, Thy will be done in earth, as it is in heaven," particularly as the lines draw out right angles at the intersections, conceptually illustrating righteousness.

Years ago I was studying quantum mechanics and was trying to gain understanding about the concept of a wavefunction. Simply put, this concept provides an illustration of the probable domain space in which a particle *may* be located. Well, the above illustration of heaven, earth, and God's people came into focus; I scribbled it on a paper and re-drew it with the parallel lines now as parallel waves connected with perpendicular lines. Instantaneously, I saw a graphic icon that could very well represent a DNA double helix. Aha! Suddenly I understood!

## LOGICS OF THE KINGDOM DYNAMICS

The Holy Spirit was revealing to me that the underlying secret was in DNA! The course of all investigation, at that moment, took a drastic turn from quantum mechanics to the genetic details of everything involved with this nucleic acid.

While there are many things to say about DNA, one thing of interest is that when a DNA molecule reproduces itself, or when messenger RNA molecules are created, information from the DNA template strand is "read" or transcribed from the end first, and this information is translated into a complementary chemical, which "begins" the construction of the new strands, whether DNA or RNA. And when all is "read" and done, the new molecule will be as the template strand, with a few exceptions in the case of RNA, which will be palindromic, read the same way from beginning to end or from end to beginning.

What was the Lord revealing to me? The DNA template strand imitates the will of God in heaven; and this will, or Word, of God in heaven (The Word being the "record bearer" as in 1 John 5:7) seeks to join with a "witness" of this will, as of the Spirit, water, and blood (mentioned in 1 John 5:8). In the same way that nucleotides bond together during transcription (the process of replication), where complementary nucleotides are created to bond to those read on the template strand, there appears to be a similar bond between the images and realities in heaven, and those that are to be created here on earth. And this understanding provides substance to the Lord's prayer: Messenger RNA is as a child of God, or Jesus, who speaks the Word (seed) from heaven on the earth and bears fruit of itself.

From this scripture we see evidence of an understanding that what is in heaven can be manifested here on earth. In heaven, there is the Word as the record bearer that declares God's will. On earth there are people (blood and water) who have been filled with the Spirit of God, and these are witnesses of God's Word, which is His will. In the same way, we may understand messenger RNA as a witness of a DNA template strand's identifying message, written out in series of nucleotides. RNA "hears" as it transcribes the message!

No less interesting is the palindromic nature of DNA, which makes each DNA strand able to be read from beginning to end or from end to beginning and have the same "message" revealed. This speaks of the eternal or timeless essence of this molecule, as well as provides a likeness to God/the Word Himself, as He declared in Revelation 1:8:

> I am alpha and Omega, the beginning and the ending, saith the Lord, which is, and which was, and which is to come, the almighty.

Isaiah 46:9-10 also says:

> Remember the former things of old: for I am God, and there is none else; I am God, and there is none like me, declaring the end from the beginning, and from ancient times the things that are not yet done, saying, My counsel shall stand, and I will do all my pleasure.

Like God Himself, DNA knows and declares the end from the beginning. In addition, each strand of DNA is like no other strand of DNA, unless it is a replication. And as God is the "I AM," DNA simply IS what it is, whatever that may turn out to be. In addition, the perpendicular pattern of connection between God's children and the Holy Spirit is mirrored by the connection between a DNA template strand and a replicating nucleic acid. Complementary base pairs (adenosine and thymine, and cytosine and guanine) are "fruitful" pairs, and are as sets of locks and keys which allow them to function as intended.

Heaven could then be understood as this realm of locks that are just waiting for the keys to open their treasures. And we have been given the keys of the kingdom (see Matthew 16:19). We have also been given all spiritual blessings in heavenly places in Christ Jesus (see Ephesians 1:3), and it is our faith that would appear to be the key to opening the locks on heaven's gates.

While I have only expressed this concept and provided this illustration to a few people over the years, I was listening to Bill Winston a while

## LOGICS OF THE KINGDOM DYNAMICS

back, and he was speaking of pulling down from the timeless domain of God's kingdom the promises that have already been provided us. Suddenly I was able to bridge my understanding with his: As we remain bound to heaven by God's Holy Spirit and His Word (just like in the process of transcription where new molecules are created), we are able to call upon everything—"All spiritual blessings in heavenly places in Christ Jesus" (Ephesians 1:3) and fully expect that they will answer our call and appear.

Being palindromic in nature, DNA does not adhere to linear time either; the beginning is the same as the end. The kingdom of God is no different, when *"the just shall live by faith"* (Habakkuk 2:4), their vision of the invisible things is as one who is declaring the end from the beginning. The reasonable assumption in this is that the end is "known" as if it had already occurred. This provides us with certainty as Jeremiah 29:11 suggests:

> For I know the thoughts that I think toward you, saith the
> Lord, thoughts of peace, and not of evil, to give you an
> expected end.

# "BE-GOTTEN"

And the Word became flesh and dwelt among us, and we beheld His glory, the glory as of the only begotten of the Father, full of grace and truth (John 1:14 NKJV).

As the Word is as the DNA pattern of God's will, if DNA were to speak, it would say, "I AM." And from this deduction, we may assume an interchangeability between God the Father and DNA. As Jesus (the Word) does what He hears the Father speak, according to John 14:31: "As the Father gave me commandment, even so I do," so, too, does a messenger RNA molecule "hear" and replicate a DNA template strand during transcription. And as the codes, or messages, are replicated, DNA would say, "*be* as I AM," in which case those new molecules would "*be*-gotten."

> In the beginning God created the heaven and the earth. And the earth was without form, and void; and darkness was upon the face of the deep. And the Spirit of God moved upon the face of the waters. And God said, Let there be light: and there was light (Genesis 1:1-3).

When God said, "let there '*be*' light," light was "*be*"-gotten: It was brought into existence by the command "BE," or, in other words, the command went forth to *exist*!

That was the beginning of the Old Testament, the inception of understanding. The beginning of the gospel of John, however, completes the picture.

## LOGICS OF THE KINGDOM DYNAMICS

> In the beginning was the Word, and the Word was with God, and the Word was God. The same was in the beginning with God. All things were made by him; and without him was not any thing made that was made. In him was life; and the life was the light of men. And the light shineth in darkness; and the darkness comprehended it not (John 1:1-5).

Commanded to *be*, Jesus, the Light of the world as well as the Word, behaved like a Messenger RNA molecule that was sent into the world to provide nourishment for it, for "It is written, Man shall not live by bread alone but by every word that proceedeth from the mouth of God" (Matthew 4:4).

The commander, God the Father, behaving like DNA, begot the Light, even as it was begotten. Light was commanded to "*be*" and the Word, as a seed, had the Light within Itself. As heat is released in the compaction of a seed during its creation, potential light energy is stored within it that will be released at the appointed time of fruition. Light, it thus appears, was in the Word, and as light is involved in the creation of matter, we understand that when we speak words, there is an inherent going forth of light energy that brings life to fruition or "*begets*" whatever has been commanded to "*be*"! Thus, the command, "*light be*" went forth, with the manifestation of Light "*gotten*" or received by all who were open to its influence, as 1 Peter 2:9 alludes:

> But you are a chosen generation, a royal priesthood, a holy nation, His own special people, that you may proclaim the praises of Him who called you out of darkness into His marvelous light.

And for those who were not? They remained in the darkness that prevented them from becoming sons of God, which reveals to us that that first command of *"light be"* was the Son of God's true conception. "In this was manifested the love of God toward us, because that God sent his only begotten Son into the world, that we might live through him"

(1 John 4:9). And as Romans 8:29 proclaims that He is the "firstborn among many brethren," we see that it is the light, which is in the Word, that is our heritage, through which we might live.

It is therefore up to us to call those things that *be* not as though they were (see Romans 4:17), and believe that we receive whatever we have spoken. As substance and evidence, our faith in our visions and expectations is truly all we need, but our faith must have been developed. We are, therefore, the ones who are responsible for our condition in life, whether our life corresponds to the darkness or to the light.

Whatever we think about and dwell upon will gain the strength of confidence in that thing's factual reality. For better or for worse, it will become the fruit of our lips, and whatever proceeds out of our mouths will pave our future paths, for "Death and life are in the power of the tongue" (Proverbs 18:21). Interestingly, both span a continuum of light frequencies, by which we are all known by God, according to Hebrews 4:13, which says,

> Neither is there any creature that is not manifest in His
> sight: but all things are naked and opened unto the eyes
> of Him with whom we have to do.

Whatever we desire to be in existence, and confidently command them to *"be," will be* according to our faith, just as a DNA template strand *will* become what it was programmed to *be*: Those who believe will then *get* the blessings and promises of God, and it can be said that the promises will have been *"be-gotten"* by those who believe.

*Watch and pray that you don't enter into temptation: the spirit indeed is willing, but the flesh is weak.*

Matthew 26:41

# ABSOLUTE TRUTH

A s a young poet, I swelled with glee at the infinite ways in which reality might be drawn. Whether it was from a wild creature's eyes or from the inner thoughts of a child, all perspectives were valid fare from which I might gain a vantage point. It was an enchanting opportunity to equalize all playing fields, with up being down, and negative space being positive. Everything was happily distilled by the one and only theory of relativity. I have always been thankful for Einstein's contributions to scientific inquiry, but while I know he tasted the sinking metallic taste of regret after inventing the atomic bomb, I wonder if he had an unction about the ominous potential of seeing things from a lens of relativity?

In the world today, we find ourselves surrounded by a dominant worldview where everything seems to be reduced in relative terms, where general reasonability appears nonexistent for many. Left is right, and right is wrong. Questions lead to disqualification of validity and worth. Anger and violence by some are tolerated and accepted, while comparatively less offensive behavior of others is scourged.

I am reminded of Acts 17:6 where Paul and Silas and many others who believed in the resurrection and power of Jesus Christ were confronted by many Jews who did not:

> But the Jews which believed not, moved with envy, took unto them certain lewd fellows of the baser sort, and gathered a company, and set all the city on an uproar, and assaulted the house of Jason, and sought to bring them out to the people. And when they found them not,

## LOGICS OF THE KINGDOM DYNAMICS

they drew Jason and certain brethren unto the rulers of the city, crying, These that have turned the world upside down are come hither also; whom Jason hath received: and these all do contrary to the decrees of Caesar, saying that there is another king, one Jesus.

In effect, those who supported the idea that the Messiah had already come in the form of Jesus were criticized for turning the worldview upside down. Those who held dearly to the status quo, however, were the ones who moved with envy—and envy led to turbulent behavior.

There appears to be a parallel occurrence today as the rules of conduct and reasonability seem to shift in a wind of popular opinion. These rules seem to be adhering to a relativism that truly is upside down, but is framing all that is righteous as that which is on its head. In relativistic terms, if you are the one who is upside down and you perceive your worldview as correct, anything that deviates from it, is itself upside down. This is a trick of our enemy because all he can do is deceive us into believing a lie. And the more these false worldviews are proclaimed as true, the less likely many will be to question them. Why? Because of the basic human need to belong.

The human need to be right (which is a values-based need) is slowly dissipating into the cultural soup of social media and the like. Very few dare to counter what their peers think or what those in the limelight pronounce. Truly left has become right, and right has become wrong. People have been able to reduce everything down to the microscopic level, where babies are simply a bunch of cells, whose coagulation does not necessarily spell the word *life*.

Now, providing us with a root understanding of relativism, an illustration depicting the notion of absolute value provides us with an overlooked truth. First, the notion of absolute value gives the same fundamental value to both five and negative five, which is Five. A shattering of the fundamental equality, however, arises when this concept is presented in a waveform construction. Illustrated as a continuous waveform with equal negative and positive amplitudes, if we were to imagine these amplitudes as a mountain and a well, the reality of these values comes to life.

If an imaginary figure were at the center of each amplitude, their realities would differ drastically: While one would be free to choose her direction and be able to see far into the distance, the other would be stuck in a hole, unable to move without assistance from an outside force. What would be required is an energy carrier that could sweep in and thrust the stranded one out, or at least a ladder that could be lowered into the well. Outside assistance would be imperative. Fortunately for us, we have such a force! And His name is the Word of God, the great I AM who declares the truth: left is *not* right; up is *not* down; and boy is *not* girl!

This illustration demonstrates how inaccurate relativity, as it has been extended to apply to our fundamental values, has been. Instead of leading us to a higher plane where our vision leads to greater potential, it has reconstructed our essential understanding of the fundamental human endowment of choice: If we are pro-life, we are against the right to choose to abort, but if we are pro-choice, we are against the right to live.

# THE COVENANT BOND

Everywhere I go, He goes, and "Every place that the sole of (my) foot shall tread upon, that has He given unto (me), as He said unto Moses" (Joshua 1:3). This is a promise of God, and I am "fully persuaded that what He has promised, He is also able to perform" (Romans 4:21). Why am I so confident? Because I know His Word, which is His will:

> And this is the confidence that we have in him, that, if we ask any thing according to his will, he heareth us: and if we know that he hear us, whatsoever we ask, we know that we have the petitions that we desired of him (1 John 5:14-15).

And

> For as the rain cometh down, and the snow from heaven, and returneth not thither, but watereth the earth, and maketh it bring forth and bud, that it may give seed to the sower, and bread to the eater: So shall my word be that goeth forth out of my mouth: it shall not return unto me void, but it shall accomplish that which I please, and it shall prosper in the thing whereto I sent it (Isaiah 55:10-11).

Years ago, at a new beach wherein a low tide had revealed an abundant supply of mussels and clams, the concept of a covenant relationship became clear. Just as the mussels and clams had become firmly adhered to a variety of rocks, wherein their bond took ardent effort to break, so,

## LOGICS OF THE KINGDOM DYNAMICS

too, is the covenant relationship between the Father and His Son, Jesus Christ, and all who accept Him as their personal Lord and Savior.

Upon deeper reflection of this relationship, I began to understand that when a smaller mussel is bound to a rock, this binding is reflective of a covenant relationship with God. Whenever a muscle is thrown or jostled in a stormy sea, the rock to which it is bound is right there, glued firmly to its neck. The rock, with its heavier weight, carries the mussel through the winds and obstacles while cushioning the mollusk from any impact. Such is our relationship with the Lord: He carries and cushions us; wherever we land, He, too, will occupy that space with us, just as Joshua 1:3 promises. This covenant bond mirrors a physical description of the strong interaction, which can only be broken by a severing blow, as with a knife or with willful disregard. However, if we continue in His Word we will be His disciples indeed, and we will know the truth, and the truth will make us free (see John 8:31-32).

While making us free, knowing the Word comes through a "bond" with the Word. And as the Old Testament definition of the word *know* alludes to the consummation of marriage, a mRNA molecule is created through a lock and key relationship (covenant bond) with a DNA template strand. It is through this process that the mRNA molecule "knows" the DNA sequence and can replicate it. And after an in-depth process of translation, the product (protein) is sent out to the body—free. Just as Jesus and His apostles were, we, too, are to be sent out to share the word: "Go into all the world and preach the gospel to every creature. He who believes and is baptized will be saved; but he who does not believe will be condemned" (Mark 16:15). And hereby our identity as one adhered to the Rock of our salvation (see Psalm 89:26) is confirmed, for wherever we go, He goes, and "Every place that the sole of (our feet) shall tread upon, that (has He) given unto (us), as (He) said unto Moses" (Joshua 1:3).

106

# IF YOU LOVE SOMEBODY, SET THEM FREE

There were many songs that held me captive when I was young; however, only a few remain firmly in my mind as having any significance to my life in general: Sting's "If You Love Somebody, Set Them Free" was one such song. Aside from its rhythm and beat, its focus on freedom drew me in every time—and I was hooked. But the notion of freedom relating to love? That, I could not understand; after all, I was in the camp of "It's better to have never fallen in love, than to fall in love and lose it."

I know I am not alone in having had the fear of loss obscure some potential blessing; but what I never knew was the magnetism of fear. As "faith is the substance of things hoped for, the evidence of things not seen" (Hebrews 11:1), fear is the substance of things dreaded and is also the evidence of things not seen. Our thoughts and our words can create and can bring our fears to pass, as Job 3:25 accurately asserts: "For the thing which I greatly feared is come upon me, and that which I was afraid of is come unto me." As electrical impulses, our thoughts, like electrons, not only harbor a light (which can create), but their movement generates an electrical current or train of thought, which can generate a magnetic field. And the magnetic field is the field in which emotions run wild—and when our emotions are in control, things rarely go well.

Fear is an emotion generating feelings of dread and thoughts that illustrate all that we do not want. But "there is no fear in love, but perfect love casts out fear, because fear hath torment" (1 John 4:18). So true

## LOGICS OF THE KINGDOM DYNAMICS

love is the antidote to the fear of loss (or any fear for that matter). But what is true love? First John 4:16 answers for "God is love," and it is His Word that has set me free from my fears, as John 8:31-32 declares:

> If ye continue in my Word, then are ye my disciples indeed; and ye shall know the truth, and the truth shall make you free.

# NUCLEAR FORGIVENESS

While we wake and sleep during the summer, winter, spring, and fall, many of us are so ensconced in our lives that we may hardly realize the many systems according to which we operate. There's the school system, the banking system, the government systems: local, state, and federal—and this doesn't even touch the behavioral, emotional, and psychological systems that become evident if we search the roots of our family's lives. These are the world's systems, and most people have been fully convinced that these are the only systems that are paramount to our existence. But there is another system: The kingdom of God system!

While many of the world's systems operate according to the same essential structure, where sowing one's money or time is rewarded later by money or some other advantage, the kingdom of God operates according to the "seedtime and harvest" system—with a premium placed on an action that the world's systems often omit. Above all, living according to this system requires faith.

In the same way that an employer will rarely dish out a pay check before a week or two of work has been completed by an employee, so, too, are we not afforded any of the benefits that God has provided *before* having done something. That doesn't mean that we must strive or work for what Jesus died to provide us, but we are to operate according to the law of faith to please God (Hebrews 11:6). And living by faith and being fully persuaded that what God has said in His Word will come to pass requires us to step out and do things with only that firm belief as a backing. In response, Hebrews chapter eleven is fully dedicated

## LOGICS OF THE KINGDOM DYNAMICS

to pointing out the works of faith that several of God's chosen people accomplished.

While there are countless scriptures relating to faithful acts that required stepping out in faith, I wanted to bring attention to one that may shed light on an old subject in a new way: Luke 6:38 (NKJV) says:

> Give and it will be given unto you: good measure, pressed down, shaken together, and running over will be put into your bosom. For with the same measure that you use, it shall be measured back to you.

We all have been reminded at one time or another that "It's better to give than to receive" and that our relationships with other people (to a large degree) rely on how much we ourselves are willing to give to those other people. However, in the matter of forgiveness, are we similarly convinced? When we for-give, what are we doing? We are giving grace to a person *before* they have shown themselves worthy of that graceful benevolence. That sounds a lot like what Jesus did for all those who accepted Him into their lives even *before* that acceptance was acknowledged by them, for:

> God so loved the world that He gave His only begotten Son that whosoever believeth in Him should not perish but have everlasting life. For God sent not His Son into the world to condemn the world; but that the world through Him might be saved (John 3:16-17).

So God gave His Son's life for the chance that those who would follow could repent of all unbelief and make the choice to accept His sacrifice in place of their own. He *"be-fore"* gave them what they would need to enter His grace: Forgiveness of sins.

And forgiveness is like a nuclear chain reaction: as we have been forgiven, we are supplied with the ability to freely give a similar grace to others. Just as a uranium atom splits as it absorbs a free neutron and releases radioactive light energy and more free neutrons, a heart that has been forgiven has been enabled to initiate a nuclear-like chain reaction. As it absorbs the loving light of forgiveness, it can release it for others to absorb. The recurring process of absorbing and releasing this light then creates a sustainable source of light energy that could potentially illuminate a city.

God has forgiven us, therefore, let us be like Him and extend the loving light of forgiveness to others, for "love never fails" (1 Corinthians 13:8).

*When it is evening you say, "it will be fair weather, for the sky is red"; and in the morning, "it will be foul weather today, for the sky is red and threatening." Hypocrites! You know how to discern the face of the sky, but you cannot discern the signs of the times."*

Matthew 16:2-3

# THE RECORD BEARER

An opening in the Supreme Court has not only brought division between political factions of late, but it has brought into focus the importance of adherence to the Constitution, as well. While one party wants someone with a strict interpretation of the Constitution, the other wants the law to be interpreted fluidly, which means they want a justice who interprets the law through a culturally-adaptive lens. In an ideal world, judges should uphold the law by consulting the Constitution and interpreting it as closely as possible—but we do not live in an ideal world. No, an ideal world was only a conception in the mind of God before He brought mankind into existence.

There is a parallel here worth mentioning: God sought righteous judges who would interpret His commandments the way He intended. His thoughts are not our thoughts and His ways are not our ways (Isaiah 55:8), however, in addition to rebellion against righteousness, bad judges came to power, just like bad judges yield power today. As the Constitution serves as the document that all judges should seek to interpret correctly, the Word of God, as Record Bearer in heavenly places, serves as the document to which all of God's children should seek to align.

It is this Word, after all, which is the final judge and juror, as John 12:48 (NKJV) declares:

> He who rejects Me, and does not receive My words, has that which judges him—the word that I have spoken will judge him in the last day.

## LOGICS OF THE KINGDOM DYNAMICS

Both the Word of God and the Constitution, therefore, serve as record bearers to which righteous judges must align themselves. Otherwise, chaos and corruption are likely to ensue.

# ENTERING INTO REST

During times when there are wars and rumors of war, pandemics, bankruptcies, stay-at-home-orders, murders, thefts, and deceptive lies being slung from every corner of the political (as well as the social) arenas, the stress that people contend with is beyond measure. But there is an antidote to stress that has been provided us as Hebrews 4:9 (NKJV) reveals:

> There remains therefore a rest to the people of God. For he who has entered His rest has himself also ceased from his works as God did from His.

This is key! When stressful events present themselves into our lives, our natural response is to scramble and feel pressured to act. But, as Psalm 46:10 declares, God's message to us is: "Be still and know that I am God." And this is a command! And in Proverbs 3:5-6, we are told to "Trust in the Lord with all your heart, and lean not on your own understanding; in all your ways acknowledge Him, and He shall direct your paths." Having faith and trust in God without relying on self is, therefore, a vital part of entering into His rest. But we must be *fully* persuaded that what He has said is truth and *will* bear fruit!

Let us look at the rest provided us: Like the rest God entered in the beginning, rather than being akin to a period of non-activity, it appears to be a time between declaration and manifestation.

> Thus the heavens and the earth were finished, and all the host of them. And on the seventh day God ended his work which he had made; and he rested on the seventh

## LOGICS OF THE KINGDOM DYNAMICS

day from all his work which he had made. And God blessed the seventh day, and sanctified it: because that in it he had rested from all his work which God created and made (Genesis 2:1-3).

Delineated in Genesis chapter one was God's calling forth of all those things that He wanted to manifest, just as Romans 4:17 declares:

(As it is written, I have made thee [Abraham] a father of many nations,) before Him whom he believed, even God, who calls those things which be not as though they were.

We see this calling forth of things that were not as though they were in Genesis chapter one when God began to say, "Let there be," and each "letting" was followed by "and God saw."

While the sixth day completed everything that God had spoken into existence, the seventh day appears to indicate a period of dormancy. This period was a duration of time where the "word seeds" that God had spoken in Genesis chapter one entered a germinating darkness. This is supported by Genesis 2:4-5:

These are the generations of the heavens and of the earth when they were created, in the day that the Lord God made the earth and the heavens, and every plant of the field **before** it was in the earth, and every herb of the field **before** it grew: for the Lord God had not caused it to rain upon the earth, and there was not a man to till the ground.

In Genesis 1:26, however, God had said, "Let us make man in our image, after our likeness, and let them have dominion." We can assume, therefore, in light of Genesis 2:5, which declares that "there was no man to till the ground," that "let us make man" in Genesis 1:26 was a "calling forth of those things that were not as though they were." Only after a "mist went up from the earth and watered the whole face of the ground" (Genesis 2:6) did God form man from "the dust of the ground and breathed into his nostrils the breath of life" (Genesis 2:7). It would, therefore, appear accurate to assume that a period of indeterminate time occurred concurrently to the mention of the seventh day in Genesis 2:2-3.

What is key to this "seventh day" time of rest, though, is the faithful force of expectation. Every word that God spoke in faith manifested because He *expected* every declaration to come to pass. This time of dormant expectation between Genesis chapter one and chapter two, however, would be most perfectly represented by pregnancy, which is accompanied not only by potential discomfort, but by *expectation*! When a woman is pregnant, it is often said that she is *expecting*. That should tell us something! If we want to give birth to any dream, we must expect for that dream to come true. Habakkuk 2:2-3 (NKJV) says it this way:

> Write the vision and make it plain on tablets, that he may run who reads it. For the vision is yet for an appointed time; but at the end it will speak, and it will not lie. Though it tarries, wait for it; because it will surely come,

It will not tarry. Anyone whose response is to run after a vision when they first lay hold upon it, exhibits their expectation of attaining the vision. When we are fully persuaded that we will attain whatever promise lies before us, it may be said that we have entered the rest of God. This same expectation of an appointed time occurs when a pregnant woman waits with faith and patience for the baby within her womb to grow to maturity. It takes faith and patience to endure this nine-month period and it takes faith and patience to enter the rest of God.

As "faith is the substance of things hoped for, the evidence of things not seen" (Hebrews 11:1), without hope, the substance of faith cannot readily develop. And it is this faithful period of expectation which appears to characterize the seventh day. After God had sowed His Word, He rested with full assurance that the word that went forth out of His mouth would not return void but would accomplish what He pleased and would prosper in the thing for which He sent it (see Isaiah 55:11).

## LOGICS OF THE KINGDOM DYNAMICS

These are the operations of God, which adhere to the process of seedtime and harvest. After the soil has been tilled and the seed has been planted in the soil, aside from making sure that enough sun and water gets to the field, the major work is that of resting in faith and having patience until sprouts of growth appear.

In our hurried and hectic lives, there are storms and obstacles that will always appear to thwart the plans and purposes of God. However, if we can remember that this seventh-day period has been blessed and is truly a testing ground for our faith, it may be a little easier to stay strong.

> Stand therefore, having your loins girt about with truth, and having on the breastplate of righteousness; and your feet shod with the preparation of the gospel of peace; above all, taking the shield of faith, wherewith ye shall be able to quench all the fiery darts of the wicked. And take the helmet of salvation, and the sword of the Spirit, which is the word of God (Ephesians 6:14-17).

# "My Policy Is the Separation between Spirit and Silicon"

On July 28, 2016, my mother and I were parked in front of a bank drive-through teller when suddenly from a pocket in my purse, we hear, "My policy is the separation between spirit and silicon." My mother looked shocked, having never actually heard my British-sounding male Siri before, and was astounded to hear this statement. I, however, was more intrigued than shocked, having become accustomed to little unexplainable treasures like these in my life.

Having made science an integral part of my investigative life for seven years at this point, since Siri's utterance, I have begun to learn about silicon, knowing that this was unlikely to have no significance. Knowing that Silicon Valley, California, is the central hub for computers and technological advancements, I could ascertain, off the bat, that *silicon* somehow represented artificial intelligence.

But there was more to Siri's statement than just the separation between Spirit and artificial intelligence. This separation speaks to all the knowledge and understanding that is obtained through the brain's activities, with the Spirit's knowledge being that which is discerned by revelation. Besides the understanding that Siri was speaking of all the devices that distract us from God, which was provided as an understanding in a posting that I found when I searched the title of this writing into my web browser, true understanding may lie in the chemical nature of silicon.

## *LOGICS OF THE KINGDOM DYNAMICS*

The eighth most common element in the universe, silicon, is distributed in sands and dusts and is used to make clay, cement, mortar, and stucco. If we were to examine the difference between Spirit and silicon with only this knowledge, what could we say? The thing that strikes me first and foremost is that while the Spirit produces the fruits of love, joy, peace, longsuffering, gentleness, goodness, faith, meekness, and temperance (Galatians 5:22-23), silicon generates no fruit. When highly purified, however, silicon is used in semiconductor electronics and is essential to anything that uses integrated circuits: Basically, all modern technological devices.

The Bible declares that we will know the signs of the times that we are living in, as everything has already been prophesied beforehand. For the last couple of years, I have been struck by how the news is accurately describing what the Bible has declared and what the Holy Spirit has placed on my heart. Ten days ago, I heard that a company was offering their employees the "opportunity" to have a chip placed beneath their skin for convenient identification purposes: They could get in the building and buy food without having to open a wallet, etc. While this technology has been around for quite some time, this "offering of the opportunity" appears to be a sure sign of end times, most likely signifying the "mark" of the beast without which "no man might buy or sell" (Revelation 13:17).

It's interesting that a chip, or *"IC"* (integrated circuit), is a set of electronic circuits composed of resistors, transistors, inductors, and diodes that are connected by conductive wires on a flat piece of semiconductor material, such as silicon. Silicon, then, acts as the body on which all these components are stored and allows them the proximity to one another that enables complex operations to be performed.

Therefore, "My policy is the separation between spirit and silicon" might allude to the eventual separation between those people who have accepted the Spirit of God to reside in them by accepting Jesus Christ as their Lord and Savior, from those people who have received the mark of the beast, or chip for the worldly conveniences that it may offer. If we look back at the parables that Jesus spoke of in the Gospels, we will see that when He spoke of the kingdom of heaven, He always mentioned a parable that either illustrated a separation between one group and another, or illustrated an object, like a field or a pearl that someone sold all to acquire.

Either way, these parables identify policy, which is described as the "wisdom involved in the management of affairs," which in these cases, reveals what will happen in the Day of Judgment:

> The Lord knows how to deliver the godly out of temptations and to reserve the unjust under punishment for the day of judgment, and especially those who walk according to the flesh in the lust of uncleanness and despise authority. They are presumptuous, self-willed. They are not afraid to speak evil of dignitaries, whereas angels, who are greater in power and might, do not bring a reviling accusation against them before the Lord. But these, like natural brute beasts made to be caught and destroyed, speak evil of the things they do not understand, and will utterly perish in their own corruption (2 Peter 2:9-12 NKJV).

It is a call to separation for those who live according to the dictates of the Spirit, who are filled therewith:

> And what agreement has the temple of God with idols? For you are the temple of the living God. As God has said: "I will dwell in them, and walk among them. I will be their God, and they shall be My people. Therefore come out from among them and be separate, says the Lord. Do not touch what is unclean, and I will receive you" (2 Corinthians 6:16-17 NKJV).

# THE PENDULUM

Tracking the progression of all trends, fluctuations are simply a part of life. From birth rates to death rates, from the financial and housing markets to all social trends, everything in life swings on a pendulum from one extreme to another. Globally speaking, in time and space, the world fluctuates between two polar points on a pendulum. It is interesting to note that when a pendulum is in a state of equilibrium, the system is at rest.

While every system is fluid and continuous, all systems seek to enter rest; they seek a point of equilibrium where all acting forces have essentially cancelled each other out. This, after all, is how major discoveries have been made. As the universe's state of balance assumes that conservation laws are at play, any observation of a lack of symmetry leads the way to discovery of the missing mass. For instance, as the annihilation of two particles as they collide creates light, any perceived charge signifies a symmetry debt of light, and this leads to the discovery of the missing charge.

The point is to highlight the conservative foundation of all universal laws. However, with the polarities of the world—light and dark; cold and heat; love and hate; and up and down—residing in a state of equilibrium in any of these domains appears impossible. Depending upon social trends and political policies as well as who is wielding power, the pendulum swings between glorifying things that reach toward the light and those things that glorify sin. These trends and policies, therefore, will either align with the Word of God, or in one way or another they will pervert it. Recently I have become aware of specific perversions,

## LOGICS OF THE KINGDOM DYNAMICS

which signify that the pendulum is closer to darkness. While all the promises of God must be received by faith to possess them, with faith as the currency, the promises are free.

So on one extreme of the pendulum we see God and Light and the freewill offering of salvation and forgiveness of sins. Bound to these gifts, however, is also a promise that "The Lord will open to you His good treasure, the heavens, to give the rain to your land in its season, and to bless all the work of your hand" (Deuteronomy 28:12 NKJV), as well as "If you can believe, all things are possible to him who believes" (Mark 9:23 NKJV).

While the Word of God offers blessings, which are free for the price of faith, we see the world's system of socialism seeking dominance as a perversion to this freewill offering by God. This system, while promising to restore a sense of equality, falls short because it is operating outside of the law of love, which is the premiere law of God, as "God is love" (1 John 4:8). This system seeks to punish *some* who have wealth, but not all—stripping them of their goods as well as freedom to choose their own paths; it emulates a dictatorship. Under the guise of a "just" system, the balancing scales are anything but balanced. As with a dictatorship, power is placed in the hands of a few at the top, while the masses remain at the bottom as subjects to their rules. It is a system where the adherents dangle the carrot of equality as well as the lure of free gifts to those who may feel a false sense of guilt or to those who are in want. All the while, it strips people of the freedom to choose—and draws them into bondage.

While all of God's gifts are given out of love for mankind, and are therefore free from bondage and are in fact provided to make people free, the gifts of the world's system are provided to enslave. Equality for all is poverty for all—except for those at the top. Inequality breeds incentive, and incentive breeds an ability to dream. Equality for all, therefore, destroys dreams. The ability to sow according to a measure within one's heart releases the invisible force of faith; and with expectation of reaping a harvest, people's lives are changed for the better.

While socialism hales itself as a just system, a truly "just" system, is one that operates according to conservation laws, where the scales are truly balanced. As human beings are endowed with free choice, a "just" law would most accurately be determined as one where the natural ability for people to choose is at its center. It is the ability to choose that creates the platform of an equal playing field. And we all have been equally provided with the ultimate choice:

> I call heaven and earth as witnesses today against you, that I have set before you life and death, blessing and cursing; therefore choose life, that both you and your descendants may live (Deuteronomy 30:19 NKJV).

The other perversion trying to gain power is the war against law and order, which stems from a perversion of the grace of God. One might say that the pendulum began at a point of grace, where God's creation of the world and mankind demanded God's ultimate favor. With grace at one end of a pendulum (conceptually that is) in the beginning, man was held above at a point with the fullest gravitational potential before falling from that grace. And when man fell from from grace, as the equilibrium point was crossed, the pendulum steadily gained potential energy in the opposite direction—toward sin, anarchy, and lawlessness. We are at this point right now.

Having crossed the equilibrium point—probably most accurately crossed in the 1960s—we have been steadily testing our own limits toward immorality once again and are beginning to see the pendulum swing back toward the equilibrium point, which if crossed could signal another gain of potential energy in the direction of grace. However, while pendulums have been the way of the world since its inception, there is an end, for Jesus said in Revelation 22:13: "I am Alpha and Omega, the beginning and the end, the first and the last." The pendulum will one day stop, and the Lord will come "with ten thousands of His saints, to execute judgment on all, to convict all who are ungodly among them of all their ungodly deeds which they have committed in an ungodly way,

# LOGICS OF THE KINGDOM DYNAMICS

and of all the harsh things which ungodly sinners have spoken against Him" (Jude 14-15 NKJV).

Years ago, I had a vision of the blessings and promises of God as waves crashing on a beautiful shoreline that were in a constant battle with an equally-strong force pulling them back into the sea. God let me know that it was in my power and authority to pull that force of undertow back! Satan is eager to steal the blessings of God from us all, and he does it often by perverting God's words so we accept his influence over our lives.

Every promise can be perverted by a simple relativistic twist so that good is evil and evil is good. *"Judge not, that ye be not judged"* (Matthew 7:1) becomes "Ignore the spiritual warnings of sin." And *"the law worketh wrath, for where no law is, there is no transgression"* (Romans 4:15) becomes, "If we can just discredit the validity of law enforcement, then we will be free from the constriction of rules." While God's message is one promoting faith and love, the perversion promotes nothing but guiltless sin, which leads to death—with no opportunity left for the pendulum to swing the other direction.

# SEEKING TO DEVOUR

With the recent devastation caused by the Coronavirus, riots, the threats of war, fires, sinkholes, the constant tearing away of the rights of the people who have chosen God's way that have occurred in the last couple of years, the reality that life and all that pertains to it is under attack is without question. I awoke this morning thinking about Matthew 11:12-13 (NKJV) which states:

> And from the days of John the Baptist until now the kingdom of heaven suffers violence, and the violent take it by force. For all the prophets and the law prophesied until John.

I was also thinking about its counterpart in Luke 16:16-17:

> The law and the prophets were until John: since that time the kingdom of God is preached, and every man presseth into it. And it is easier for heaven and earth to pass, than one tittle of the law to fail..

Alluding to the same idea, we can reasonably conclude from these scriptures that the kingdom suffering violence has something to do with the kingdom of God being preached.

While I have heard many speak about Matthew 11:12 and use it as a point of reference that validates the body of Christ being bolder in their approach toward the things of God, I have come to understand these scriptures in a new light. The advent of John the Baptist preaching in the wilderness of Judea and saying, "Repent, for the kingdom of heaven

## *LOGICS OF THE KINGDOM DYNAMICS*

is at hand" (Matthew 3:2 NKJV), marked the beginning of all prophetic fulfillment: "For this is he that was spoken of by the prophet Esaias, saying, The voice of one crying in the wilderness, Prepare ye the way of the Lord, make his paths straight" (Matthew 3:3). Much like Jesus himself, John the Baptist was like a seed bursting forth out of the ground which had existed up until that point only in prophecy. The fruit of the Old Testament's laws and prophets had finally come to bear!

As seeds in the darkness remain dormant until the moment they split in the initiation of new growth and reach toward the light, so, too, was the coming of John the Baptist. The law and the prophets had prophesied about him, and the waiting period in proverbial "darkness" constituted another "seventh day" period of germinating darkness. And just as it is with the fulfillment of a pregnancy, the manifestation of the promise is a violent breakthrough event. Such was the significance of the kingdom of heaven suffering violence at John's appearance. This violence was a direct response to the kingdom of God being preached!

While preaching the Word was apparently a foolish endeavor, according to 1 Corinthians 1:21: "For since, in the wisdom of God, the world through wisdom did not know God, it pleased God through the foolishness of the message preached to save those who believe." And the eventual end of that preaching—that people would hear and believe the Word spoken—was the very substance over which the violence occurred, and is still occurring! Essentially, it is a fight between life and death. John 6:63 declares the battleground: "It is the Spirit who gives life; the flesh profits nothing. The words that I speak to you are spirit, and they are life."

> But the natural man does not receive the things of the
> Spirit of God, for they are foolishness to him; nor can
> he know them, because they are spiritually discerned
> (1 Corinthians 2:14 NKJV).

Understanding spiritual things, therefore, requires revelation knowledge, which is like a flash of lightning that breaks through the darkness. Isaiah 9:2 alludes to this revelatory moment: "The people that walked in darkness have seen a great light: they that dwell in the land of the shadow of death, upon them hath the light shined." And this light was none other than Jesus, the Word of God. Without God's intervention, therefore, revelation will not come, for He is that Light that breaks through the darkness.

The violent taking it by force, therefore, alludes to the fight against the manifestation of the promises of God. Dark forces, being anti-revelation, seek to deceive the minds of those who might be exposed to the light of revelation—they seek to destroy all prophetic fulfillment.

And herein is the revelation confirmed:

> And there appeared a great wonder in heaven; a woman clothed with the sun, and the moon under her feet, and upon her head a crown of twelve stars: and she being with child cried, travailing in birth, and pained to be delivered. And there appeared another wonder in heaven; and behold a great red dragon, having seven heads and ten horns, and seven crowns upon his heads. And his tail drew the third part of the stars of heaven, and did cast them to the earth: and the dragon stood before the woman which was ready to be delivered, for to devour her child as soon as it was born (Revelation 12:1-4).

Satan's time of ruling on this earth is ending—and he is on a rampage to devour the promises of God. Like the afore-mentioned dragon, he is poised before every one of us, ready to give a partial-birth abortion to every promise that God has given us, as well as all those things that we have been called forth in faith, believed to be received.

> Be sober, be vigilant; because your adversary the devil, as a roaring lion, walketh about, seeking whom he may devour: Whom resist stedfast in the faith, knowing that the same afflictions are accomplished in your brethren that are in the world (1 Peter 5:8-9).

# MAN versus MEN

Years ago, when media attention was placed on MS-13 (an international criminal gang) in response to a recent killing of two young female students by gang members, President Donald Trump referred to them as "animals." As you can imagine, the anti-Trump media went wild! A member of congress even reframed the members' identity as "God's Children." And commercials touted the gang's members as students whose differences, such as tattoo-laden faces, should be embraced. Yes, we have been commanded to "judge not, that ye be not judged" (Matthew 7:1); we have been commanded to love, bless, do good to, and pray for those who come against us (see Matthew 5:44). However, we have also been called to discern between good and bad.

Malachi 3:16-18 (NKJV) sheds light on this calling:

> Then those who feared the Lord spoke to one another, and the Lord listened and heard them; so a book of remembrance was written before Him for those who fear the Lord and who meditate on His name. "They shall be Mine," says the Lord of hosts, "On the day that I make them My jewels. And I will spare them as a man spares his own son who serves him." Then you shall again discern between the righteous and the wicked, between one who serves God and one who does not serve Him."

President Trump's reference to MS-13 gang members as "animals," while being potentially offensive, could have been an accurate description. After all, from a macro-perspective we are animals, not

*LOGICS OF THE KINGDOM DYNAMICS*

vegetables! Seriously though, when man was created in the image and likeness of God (see Genesis 1:26), he apparently existed in a germinating darkness, or a "seventh day" domain space. However, when "a mist went up from the earth and watered the whole face of the ground" (Genesis 2:6 NKJV), which could be understood as a "former rain" (Joel 2:23), the moment of fruition of everything that had been spoken came to pass. Then "the Lord God formed (made) man of the dust of the ground, and breathed into his nostrils the breath of life; and man became a living being" (Genesis 2:7 NKJV). At this point, this man who had been conceived in God's mind was manifested in the flesh.

In Genesis 1:24, which illustrates a manifestation of living creatures, we see that the earth appears to have already been filled with life. However, as an indefinite amount of time passed between Genesis chapter one and Genesis chapter two, it remains uncertain how long these other creatures had inhabited the earth. However, what appears definite is that God wanted a man who would have His own heart, mind, and soul. This man, Adam, would be as a son to Him, and to whom He could be a Father. When God formed man from the dust of the earth and breathed into him the breath of life, His vision became a reality.

Adam talked with God and took instruction from Him, for "The Lord God commanded the man, saying, 'Of every tree of the garden you may freely eat; but of the tree of the knowledge of good and evil you shall not eat, for in the day that you eat of it you shall surely die'" (Genesis 2:17 NKJV). So Adam showed himself to be the first oracle of God. Essentially, this man "Adam" was created to be in a covenant (lock and key) relationship with God, to whom God could commit "the oracles of God" (Romans 3:2). The identity of this man who was created in the image and likeness of God, as is confirmed in Romans 3:1-2 (NKJV), appears to have been the Jew:

> What advantage then has the Jew, or what is the profit of circumcision? Much in every way! Chiefly because to them were committed the oracles of God.

When God breathed the breath of life into Adam, not only was Adam made a living being, but he had been given authority to name things. This means that Adam could act like God, who calls "those things that be not as though they were" (Romans 4:17). Here is how Adam responded to this authority:

> Out of the ground the Lord God formed every beast of the field and every bird of the air, and brought them to Adam to see what he would call them. And whatever Adam called each living creature, that was its name (Genesis 2:19 NKJV).

Adam had been empowered to speak words and make decrees, and he could use them to manifest whatever was in his mind just as God had. And as Adam was the first Man, a Jew, with whom God could impart His blessings, we see that Jesus is referred to as the second or last Adam:

> And so it is written, "The first man Adam became a living being." The last Adam became a life-giving spirit. However, the spiritual is not first, but the natural, and afterward the spiritual. The first man was of the earth, made of dust; the second Man is the Lord from heaven (1 Corinthians 15:45-47 NKJV).

The first Adam, a Jew, and the second Adam, Jesus (also a Jew), had been separated out from the other creatures, and both had been endowed with creative power. While Jesus has been "given...the name, which is above every name" (Philippians 2:9 NKJV), Adam had been given the power to name every living thing. So what were these other creatures for which Adam provided names? Genesis 1:24-25 provides evidence of their conception:

> And God said, Let the earth bring forth the living creature after his kind, cattle, and creeping thing, and beast of the earth after his kind: and it was so. And God made the beast of the earth after his kind, and cattle after

# LOGICS OF THE KINGDOM DYNAMICS

their kind, and every thing that creepeth upon the earth
after his kind: and God saw that it was good.

While we know very little about these "beasts of the earth," other
than our assumption that they were animals of some sort, in speaking
about people whose hearts do not emulate God's, 2 Peter 2:12 declares:

But these, as natural brute beasts, made to be taken and
destroyed, speak evil of the things that they understand
not; and shall utterly perish in their own corruption.

In stark contrast, we see that Abraham's nephew, Lot, who was of
the lineage of Adam, a Jew, while living amongst sinners in Sodom and
Gomorrah, was referred to as "that righteous man dwelling among them,
in seeing and hearing, vexed his righteous soul from day to day with
their unlawful deeds" (2 Peter 2:8). I make a distinction here between
*man* and *men*. If we go back to Genesis chapter four, we see that after
Adam's son Cain killed his brother Abel, he went off and dwelt in the
land of Nod, east of Eden. And it was in Nod that Cain apparently found
a wife, who obviously had a different lineage than his.

After Cain had killed Abel, Adam and his wife had another son
named Seth. Seth had a son named Enoch, and then the descendants of
Adam began to call on the name of the Lord. This became the lineage
of the sons of God. But Genesis 6:1-2 reveals how this lineage became
corrupted:

And it came to pass, when men began to multiply on the
face of the earth, and daughters were born unto them, that
the sons of God saw the daughters of men that they were
fair; and they took them wives of all which they chose.

This should lend pause to all who declare that Adam was the first
created *man*—he was not; but was rather the first Jew with whom God
would be in covenant. He was also the first man who was given the same
authority as God to call "those things which be not a though they were"
(Romans 4:17). He was the first of a chosen people who would glorify

God by adhering to His commandments. So we see here a definitive distinction between one creation who are called the sons of God, who came from the "man" in the image of God, and another creature, who are referred to as sons and daughters of "men." Of interest, as well, is the fact that while Jesus was on earth, He was referred to as the Son of *man*—not as the Son of *men*.

*Man* refers to a peculiar person, a new creature, a prototype of what is available to everyone who accepts Jesus Christ, the second Adam, into their hearts! In addition, "man after God's own image" appears to refer definitively to the Israelite, which provides new insight into the factions of mankind within the world today. This also provides a distinction between "created" and "made." "Man" was created while "men" were made. Things that are created adhere to eternal spiritual principals, while things that are made are temporal. Hebrews 12:25-27 (NKJV) provides this clarification:

> Him who speaks from heaven, whose voice then shook the earth; but now He has promised, saying, "Yet once more I shake not only the earth, but also heaven." Now this, "Yet once more" indicates the removal of those things that are being shaken, as of things that are made, so that things which cannot be shaken may remain.

Those who call upon God and His Son, Jesus, have assumed a "created" nature and will not be readily shaken as those "men" who remain in a temporal or "made" state.

When we hear reference to some "men" as "animals," therefore, it may be that they are simply brute beasts whose mere existence is fitting for a pre-ordained end. And as the first man, Adam, had been given authority to name "every beast of the field and every bird of the air" (Genesis 2:19), these were simply among the "called." But we all have been called to a certain end and we all have been given a choice:

> I call heaven and earth as witnesses today against you, that I have set before you life and death, blessing and

## LOGICS OF THE KINGDOM DYNAMICS

cursing; therefore choose life, that both you and your descendants may live ( Deuteronomy 30:19 NKJV).

# Author's Note

I have always been enthralled by bridges; not as a counter to the building of walls, but more as a construct that connects disparate themes. As a poet, I loved being slapped with an image that was wrapped in a similar blanket to another, with its ends fraying in hope that one would mend its ways. In response, I always gathered a needle and thread and went to work.

And as it is with poetry, so it is in bridging science with the Bible; both have required a good amount of faith as well as patience. They also both require meditation on the finer points of life as well as listening until the light of revelation settles in the heart and in the mind. Then and only then can words articulate an apt reflection. "(Calling) those things which be not as though they were" (Romans 4:17), I declare that every person who reads this book has an understanding heart in which the "seeds" of the Word will be planted and will bear fruit!

> But he who received seed on the good ground is he who hears the word and understands it, who indeed bears fruit and produces: some a hundredfold, some sixty, some thirty (Matthew 13:23 NKJV).

> But what does it say? "The word is near you, in your mouth and in your heart" (that is, the word of faith which we preach): that if you confess with your mouth the Lord Jesus and believe in your heart that God has raised Him from the dead, you will be saved. For with the heart one believes unto righteousness, and with the mouth confession is made unto salvation. (Romans 10:8-10 NKJV).

# Acknowledgements

Above all, I must thank God for putting a fire beneath me to reach out to Andy Sanders, with the quiet reminder that, "You have not, because you ask not!"

I thank Andy for encouraging me to reach out to his wife, Cathy Sanders, whose dedicated work in putting this book together along with her patience and professionalism has been inspiring. Thank you!

I also want to thank Noah Fortinberry for his constant encouragement and support, as well as for his contribution in designing the cover of this book.

And I must thank Joseph Delillo for being a source of encouragement and a listening ear when these treatises were first written, and with whom long discussions ensued regarding the science of the Bible. Thank you, Joe!

# INDEX

Acceleration: 15

Attraction: 21

Bond: 18, 92

Boson: 28

Breakthrough: 130

Chemical: 17-18, 92, 121

Constructive Interference: 50

Decay: 15, 28

Density: 16, 17

Electrons: 21-22, 37, 57, 59-60, 75-76, 107

Emotional Baggage: 16-17

Expectation: 119, 126

Fission: 68

Grave: 43

Gravitational Potential Energy: 17, 39

Gravity: 31-33

Halide Crystals: 75

Hypotenuse: 43

Impulses: 57-58, 75, 107

Keys: 93

Magnetism: 23, 61, 107

Measurements: 12, 15-16, 45

Mountain: 5, 37, 68, 84, 88, 103

## LOGICS OF THE KINGDOM DYNAMICS

Mussel: 106

Palindromic: 92

Photons: 75

Photographic: 75

Proton: 21, 28

Puzzle: 46-47, 55

Quantum Mechanics: 45-46, 91

Quarks: 28

Receipt: 11, 84

Renewing: 15, 33, 73

Repentance: 24, 55-56, 61

Resistance: 17, 25, 35-37, 39-40

Restoring: 15, 33

Unlimited Potential: 16-17

Violence: 67, 69-71, 101, 129

Volume: 16

Wavefunction: 45-47, 91

# About the Author

Heather Fortinberry is a bridge builder, drawing connections between scientific, psychological, and social inquiry and the Word of God, as revealed in scripture. At just seventeen years of age she had been in a coma. Since that experience, Heather spent years studying scientific processes and the fundamental forces of the universe.

Using the Bible as her source of strength combined with her extensive knowledge, Heather published a developed theory of the universe in 2014 entitled *Logics of the Kingdom: A Scientific Analysis of the Word of God*. She describes physical phenomena as manifestations of spiritual principles and proves how the fundamental theory lays a foundation upon which understanding can be built.

Heather is a life coach whose primary motivation is to provide people with a sound explanation of the world around them with the hope that knowledge and understanding will be gained and that they might fulfill their potential. Heather has a master's degree in psychology and taught for ten years at Brook's Institute in Ventura, California.

For more information, please contact hfortinberrylifecoaching.com or logicsofthekingdom.com.

Made in the USA
Coppell, TX
27 October 2020